MA
YO

D0825908

LEADER'S GUIDE

STEPHEN B. DOUGLASS

CAMPUS CRUSADE FOR CHRIST INTERNATIONAL

MANAGING YOURSELF LEADER'S GUIDE

Copyright © 1980 by Stephen B. Douglass

A Campus Crusade for Christ Book

published by:

Here's Life Publishers, Inc.

2700 Little Mountain Drive

P.O. Box 1576

San Bernardino, California 92402

Printed in the United States of America

ISBN 0-918956-69-2
40-089-5

CONTENTS

Introduction

"We are all faced with many, many demands on our time. The boss wants that project now, the telephone rings throughout the day, people drop in unexpectedly at home or at the office, the television tells us that we should be enjoying all sorts of pleasant activities, the pastor encourages us to become more involved, the kids want to play baseball, the family wants to go out to eat. Sound familiar?

"There are so many choices. And yet we alone must decide what to do next. We may decide in response to circumstances. We may decide in response to obligations we feel to people. We may decide out of habit or tradition. We may do only what we enjoy. Or, we may make purposeful, God-inspired decisions. One way or the other, we decide.

"In this world of high pace and complexity it would certainly help us if we learned to be good managers or stewards of ourselves" (from Chapter 1 of *Managing Yourself*).

The objective of this *Leader's Guide* is to make it possible for you to help people be good managers of themselves. It will assist you in guiding people through the book *Managing Yourself*. The content is well suited to people high school age or older. It is equally applicable to men and to women.

The basic objectives of *Managing Yourself* are to cause people to learn and apply the following:

1. How to set objectives for their lives.
2. How to determine the areas of their lives in which they most want to see accomplishment or improvement in the next six to twelve months; and then how to determine the best way to achieve those accomplishments or improvements.
3. How to schedule each day in order to accomplish their highest priorities.
4. How to follow through on what they plan for each day.

5. How to walk more closely with God and listen to His directions.

This *Leader's Guide* can be used in several different formats of study groups or classes. Its basic structure provides for 13 sessions of 45 minutes duration. This should fit comfortably into a Sunday school class situation or a day or evening weekly study situation.

If a briefer coverage is desired, the content can be trimmed to ten or even five 45-minute sessions. This allows for use in Campus Crusade for Christ Leadership Training Classes, in weekend retreats, and in one-or-two-week special evening emphases in churches or other settings.

On the college campuses this content has been used to attract non-Christian student leaders due to their special interest in managing themselves better. A one-hour, guest-lecturer format and a three-to-four-hour Saturday morning mini-retreat format seem to be the best approaches for this purpose. Special sensitivity is needed in these situations in working in evangelistic and other biblical material. Many non-Christians in the broader adult community have a similar interest in making the most of their time.

For more in-depth study, additional material is suggested for every session which would enable each session to extend for up to as much as two hours. This would provide suitable curriculum for lengthier weekly studies and Bible institute formats. In the more leisurely context of a family devotions, the basic session content, plus the suggested additional material, could provide fruitful discussion and project topics for a half year and beyond.

HOW TO SUCCEED AS THE LEADER OF YOUR GROUP

Pray

As with anything else in the Christian life, the best way for you to succeed as the leader of the group is to trust God. One of the best means and evidences of trusting God is to pray. Why not pause even now and ask God for His perfect wisdom as you consider how to handle this group for His maximum glory. Ask Him to give you a vision of people's lives being positively changed. Continue to pray as you prepare for and conduct the sessions that follow.

Seek to find other people who will commit themselves to pray for you and for this particular study group or class. Give them some of the details of what will be covered each week, which will enable them to pray specifically.

As soon as you have the opportunity to do so, communicate with the members of your group. Encourage them to start praying that God will enable you to make the content clear and palatable to them. Encourage them also to pray for themselves, that their hearts will be prepared to receive knowledge on this subject, and that they will apply it. If your group is a Sunday school class, you could conveniently initiate such spiritual preparation several weeks before your sessions begin.

As you contact your members—and during the first session— suggest that they team with others as prayer partners to pray daily for themselves, each other and you.

Establish Objectives

Most people are more motivated and effective when they know where they are going. What do you hope to accomplish by teaching this study? As you think of the members of your group, what kind of progress do you pray they will see in their lives? Do they need more overall direction from God? Do they need to make better use of their time? Do they need more personal motivation and discipline? Since the leader usually learns even more than the group members in a study, what potential benefits to your life do you see from doing this study? Take a moment to write the answers to these questions in the space provided on the next page.

MY OBJECTIVES FOR THIS STUDY

Read through these objectives each week as you prepare for the study. Pray about them. Ask God for a real vision and excitement toward their accomplishment. Add to them as more thoughts occur.

Adapt the Contents to Your Group Situation

If you are planning to cover this material in 13, 45-minute sessions, I recommend that you follow the basic structure outlined in each session under the heading "Conducting the Session." If, for some reason, you must shorten the series to the equivalent of five, 45-minute sessions, I recommend that you simply cover Sessions 1-5. These sessions will provide you with a basic coverage of the subject.

If you plan more than five, but fewer than 13, 45-minute sessions, I recommend covering Sessions 1-5, then selecting from the remaining sessions ones of special interest to your group. For example, if your group has a particular interest in what the Bible has to say about handling money, be sure to include Session 11 (about the financial area of life). In the event you eliminate some sessions, be sure to correlate your assignments for the next session properly.

If the time available for your session is to exceed 45 minutes, see the session section titled "Additional Material" for ideas on how to extend the length of your session.

Session 1 includes the content of Chapter 2 of *Managing Yourself*, "The Spiritual Prerequisites to Managing Yourself." This material is important to the overall subject. But, depending upon the expectations of your group members and their previous exposure to the Transferable Concepts of Campus Crusade for Christ, you may choose to cover the material toward the end of the study.

In a Friday night and all day Saturday retreat situation, you may want to cover Sessions 1-5 to give the basic content on the subject. On subsequent retreats, or as an alternative to the above, you may want to cover two or three of Sessions 6-12, adding some or all of the "Additional Material" sections to provide for more in-depth discussion and other involvement.

Start Managing Yourself

It is important to try to model the truths we teach. Jesus was a good teacher partly because His actions were good examples of what He taught. You may already be pretty good at managing yourself. But if you are not, there is still time to take a few steps out in front of the members of your study. As soon as you can, read through *Managing Yourself* and try out some of the key concepts. You may not have perfected things by the time you cover it with the group, but at least you will not seem to be teaching theory.

Arrange for the Right Setting

For some reason, informality seems an aid in learning. So, if you have a choice in the matter, try to pick an informal room that isn't so comfortable as to encourage people to fall asleep. If you're stuck with a particular room, that is more formal, at least arrange the chairs in a circle or some other pattern which will be conducive to discussion. If the group is small enough, sit down in the circle as a member of the group with your visual aids beside you. Of course, if the group is larger, you will have to stand up just to be seen. Make sure that the air conditioning or heating is working properly. Nothing can make a lesson stuffier than a stuffy room.

Recruit the People

Even if you regularly teach a class to which you will now be teaching this material, plan on making a concerted effort to motivate the people to attend. Describe the benefits they can reasonably expect to receive from being a part of this study group. Encourage each person to bring a friend if that is appropriate. If someone other than you is responsible for promotion, make sure that they have a clear idea concerning the purposes, content and benefits of the study. The back cover of *Managing Yourself* will provide you with a few paragraphs that may be helpful in your promotion.

If you feel God is motivating you to teach this subject even though you don't regularly teach a group, you might consider forming a new group. Make a list of those who might be interested in the subject. If, for example, you are a student in a dormitory at a college, there are probably other students in your dorm who would be interested.

Place a notice on the bulletin board featuring the front and back covers of *Managing Yourself* along with your own words of encouragement. Be sure to include the date, time and location. Don't overlook word-of-mouth advertising among your friends. It is one of the most effective methods of attracting a group for any purpose.

If you intend to start a Bible study in your home, one of the most convenient groups to invite would be your neighbors.

Another would be friends from church, work or clubs of which you are a part.

Obtain the Needed Materials

First: Obtain an adequate supply of the text, *Managing Yourself*. They can be purchased through your local bookstore or by ordering directly from Here's Life Publishers, 2700 Little Mountain Drive, San Bernardino, CA 92405. Be sure to allow enough time for order processing and delivery that will enable you to have the books on hand about one week before your first group meeting. It is very important that you distribute the texts to your group members, along with the first assignment before the first session. For instance, if you plan the normal suggested 13-week schedule, the assignment will be to read the first two chapters of the book before coming to the first session.

Second: Remind your students to bring writing materials and a Bible.

Third: You will note that I have suggested one major priority area of change be highlighted in each session. In the course of the entire study this procedure will generate many potential areas for each member. Realistically each person should probably seek to achieve change in only one or two areas at a time. Therefore, it is wise for each group member to write down and save the remaining potential areas of change for his or her future planning. If you want to encourage this, you may want to arrange for notebooks or folders for the group members.

Fourth: I highly recommend the use of a visual aid of some sort in your presentations. You will notice that some are mentioned specifically in the sessions. Your choice may be a chalk board, an overhead projector or a flip chart. Use whatever you feel most comfortable with, realizing that visual aids are excellent tools for clarifying content and focusing discussion.

Fifth: Decide what kind of refreshments (if any) are going to be served and arrange for them.

Sixth: Make arrangements for defraying the expenses for the texts, refreshments, etc.

Prepare for Each Session

Turn to Session 2 in your Leader's Guide and refer to that section as you continue reading this introductory section. Note the four major headings in each session's material. First is "Session Objectives," which tell you what you should try to accomplish in that session. Second is "Leader Preparation," which is a check list of specific things you should do or think about or pray about before the session begins. Third is "Conducting the Session," which gives you a proposed schedule of what you should cover in the session itself, including extensive guidelines that will help you accomplish each session's objectives. Fourth is "Additional Material," which provides additional content ideas to be used either for extended sessions or for substitutions for other session items.

Let's examine these sections in detail in order to maximize their benefits to you and your group.

A. *Session Objectives.* These are what each group member should try to accomplish in the session. Everything you do and say in your group should seek to contribute toward the realization of these objectives. Study these objectives carefully. Pray about them. Keep them before you as you prepare for the session.

B. *Leader Preparation.* The following is a further explanation of the leader preparation checklist you will find in each session.

1. *Pray for God's wisdom in your preparation.* Pray specifically for the objectives. Pray that God will give you special wisdom to determine if any customizing of the material is needed, and if so, how to do it. Pray for sensitivity to the special needs of your group members in this particular area of study.

2. *Read and meditate on the session objectives listed above.*

3. *Read carefully the chapters that are going to be covered in this particular session at least once.* Obviously you'll be much better prepared to teach if you read these chapters several times. In some cases you will be given an additional chapter to use as background material for the session.

4. *Read the rest of the session's material several times.* Since you will be teaching from this material, it is essential that you make it your own, annotating and customizing it to fit both your and the group's needs. Become very familiar with the flow of the content.

5. *Determine what overall customizing of the material you plan to do.* Some factors to consider are the length of time you have for the session, the special needs and interests of the group, and the choices you are presented in the "Conducting the Session" and "Additional Material" section. (If you want to use optional, assigned presentations by the group members, it may be wise to look through the session materials for the session after the one for which you are preparing. This will enable you to make the assignment on a timely basis.)

Extra space is provided between the paragraphs and in the margins of the "Conducting the Sessions" section which will allow you to make note of any additional information you plan to incorporate. If you have more time than the suggested schedule, you may want to include one or more of the ideas from the "Additional Material" section, especially ones you feel would be particularly appropriate for your group.

Most people find it helpful to estimate how much time they should spend on each item so they can pace themselves and thus cover all of the ground adequately as they teach the group. To assist you in this a suggested time schedule is provided in the left margins of the "Conducting the Session" section. These suggested times add up to 45 minutes. (Notice that in the proposed schedule for the first session, 15 additional minutes have been allowed for getting acquainted. If you don't need this time, simply move ahead on your schedule.) As you move through the material proposed for the various time slots you will see that you will have to do so rather quickly if you are to cover the material in the allotted time. If for you or your group you find that it would be unrealistic to cover so much material in the scheduled time, you will need to adjust accordingly. This can be done either by deleting some of the material or by extending the session.

6. *Prepare for personal illustrations, examples and group member assignments.* Read carefully through the "Conducting the Session" section. Take note of the personal illustrations that are called for. Since personal illustrations communicate so well, look for other spots in the session where you can mention some. Be very familiar with any examples, role plays or other items that will require special preparation. If you are going to make assignments to the group members for the next session be familiar enough with the text so you can answer any questions that might be asked.

7. *Arrange for final needs related to facilities, visuals, refreshments, etc.* Some other needs might include nametags, note paper, pencils, reminders to the host and group members, special materials in response to a question from last week, and some extra books and other materials for guests.

   The use of the symbol Ⓥ in the "Conducting the Session" sections of this *Leader's Guide* indicates a good place to consider using a chalk board or other visual aid to stimulate the learning process.

8. *Pray that special needs will be met in the lives of the group members and that the session objectives will*

be accomplished. With each session spend time in prayer *after your preparation, praying* for the specific things that you are hoping will happen as a result of this session. Pray for each of the various segments of your presentation. Pray for the people in your group who have special needs in this area. Pray for their participation and their willingness and openness to be ministered to during the course of this session. Pray that you will conduct the session well.

C. *Conducting the Session.* This section gives you suggestions on how to cover the material in a 45-minute time period. Although there are a number of variations, the proposed flow of content normally includes the items listed below. (Refer to the "Become as Skilled as Possible in Leading the Group" section of this Introduction for further help in how to do some of the following activities.)

1. Begin with prayer.
2. Review the concepts covered in the last sessions and the application made by the group members.
3. Read the objectives of the session.
4. Lead into the subject by highlighting a portion of *Managing Yourself,* asking a key thought question, etc.
5. Cover the main body of the content by highlighting more of the text, asking further discussion questions, studying passages of Scripture, considering case studies, working through examples, hearing assigned reports from group members, brainstorming, and considering real life situations faced by members of the group.
6. Ask each group member to select from the session's content the greatest need for application he or she has in his or her life. Have each group member think about how to implement that application.
7. Give the reading and other assignments for the next session.
8. Close in prayer.

D. *Additional Material.* Although the material outlined under "Conducting the Session" should normally be sufficient to accomplish all the session objectives, this section includes additional content ideas that can be added to the session or used as additional homework assignments, or substituted for existing "Conducting the Session" material. If you and the group members desire a more in-depth coverage of the material, the ideas in this section should be very helpful. This additional material normally includes several of the following:

1. Additional and related discussion questions and other involvement ideas.
2. Additional thought questions to raise issues that could be helpful to the group's understanding of the subject.
3. Specific homework assignments.
4. A Scripture passage related to the session that the group could memorize.
5. One or more creative, "different" approaches such as dramatic reading, singing, a debate, a role play or other teaching techniques that would dramatize the lesson.

The above explanation should give you a greater understanding of each of the four parts of the session materials and how you can use them to maximize your effectiveness as the leader of the study.

Become as Skilled as Possible in Leading the Group

A good share of teaching today, even in small groups, is still done by the lecture method. It is a good method for conveying a lot of information efficiently. In some parts of the sessions you are encouraged to lecture for a short period of time. However, since the text carries the main load of conveying conceptual information, you are also encouraged to use other teaching techniques in conducting the sessions. The following thoughts should help you use these other techniques.

A. *ENCOURAGING DISCUSSION*

Meaningful discussion normally produces an excellent learning situation. The purpose of good discussion is to involve the group member, thus causing him to learn more.

Your role in discussion is to stimulate, but not dominate the conversation. As the discussion picks up, cut back on the number of comments you make. Reward group members' good comments both by recording them on your visual and by acknowledging them as good points. Keep the discussion moving by saying, "Anyone else have a comment on that?"

Don't be concerned about short periods of silence, especially when the discussion question has just been asked and the group is pondering the answer. If the silence continues, ask the question another way. Try to ask open-ended questions which cause the group member to see that there are many answers he could volunteer versus one "right" answer you want him to say. "What are some reasons Christians should manage ourselves?" is an example of an open ended question. Also ask wide-open questions such as, "Any other thoughts?"

Make an effort to arouse interest; then pass the ball to others in the group. Sometimes even a little bit of controversy is helpful. For example, if one of your group members who is not easily threatened makes a statement or comment with "holes" in it, you can stir up healthy discussion or controversy by asking the group, "Do all of you agree with that?"

Sometimes discussion seems to drag in response to several questions in a row. Often the problem is that most of the people in the group are not really thinking very hard about the answers, figuring someone else will speak up. This is an ideal time to use a technique called "buzz group" or "breakdown group." The purpose of this method is to involve more people actively. You begin by dividing your total group into small groups, with even as few as two or three people per small group. Next assign the question(s) to be discussed. In the small groups every person will have opportunity and responsibility to contribute to keep the discussion flowing. After the small groups have discussed the question(s) for awhile, you can re-form the larger group and have a lively discussion of the total findings of the small groups. A key to this method is to give the small groups a very easily defined question or topic to explore. Otherwise they may become confused and stalled.

In a discussion setting, if you want to generate a large number of creative thoughts on a subject in a short time, use the technique called "brainstorming." In brainstorming you pose the problem or thought question to the group and ask them to mention *any* solution or other thought that might contribute. There is just one ground rule: No one can criticize or make fun of what anyone else has offered. Ask people to reserve their judgment on the "rightness" of an idea until later. Be sure to use a chalkboard or other visual to record ideas as they come. This helps cause further ideas to follow from ones that have already been expressed.

B. *LEADING THE MEETING*

Even though you want to encourage the maximum amount of group involvement, you must keep in mind the objectives of your overall session. Try to keep approximately on schedule. One of the best ways to do this is to give the group some idea of how much time you will be spending on the point. Then most group members will tend to limit themselves. You may have to learn how to handle certain over-talkative people. A comment such as, "That's good, Sam. What do the rest of you think?" usually works by rerouting the question to others.

Even though you want to be somewhat clock-conscious, you must also be *very* sensitive to the needs of the group. If the group members are really coming to grips with a significant point for them, it is usually wise to let the discussion run for awhile and skip something else later. The group members will feel that they have had good leadership if they sense they have been allowed to discuss meaningfully important issues without getting stuck on issues that didn't seem worth their time. This will require wise judgment on your part. It will also require an equitable balance between your being assertive with the group at times and being willing to fade into the background at other times.

Occasionally you will have to step in and clarify what is meant by a question. Sometimes you will have to split the issue when a discussion becomes too complicated, and suggest the group discuss it in two parts. When you need to move in and grab the attention of the group, you can often do this just by approaching the visual and by summarizing the matter at hand. Or you can wait for an opening and, by using an appropriate question or comment, easily gain control of the discussion.

Remember, you know better than anyone else what you hope to accomplish from the session. If the ultimate objective is to be realized in your group members' lives, it's your obligation to see that they have the opportunity to reach that point.

C. *LEARNING HOW TO INVOLVE PEOPLE*

In the course of conducting the sessions, you will be asking people to comment on a point, participate in a role play, relate their experiences, or prepare mini-lectures. As you do this, keep the following guidelines in mind:

1. Make sure the persons called upon are capable of doing whatever you are asking them to do without being unduly threatened. If you have some doubts, be sure to ask them in private ahead of time to see if they would mind being involved in this way.
2. Tell them exactly what you have in mind and offer to help them accomplish the assigned task.

3. Encourage them—when you give them the assignment, while they are performing and after they've accomplished the assignment. Remember, it's very difficult to encourage a person *too much*.

4. Look for and employ people who have special talents. For example, some people have special talents in writing or drama or singing or humor. When it comes time to do a role play, it would probably be best done by a person with skill in drama or humor. Such persons can be a real blessing to the group.

D. *TRYING OTHER HELPFUL CONCEPTS AND TECHNIQUES*

1. Learn more about how to use different teaching techniques. One excellent book on this subject is *24 Ways to Improve Your Teaching* by Kenneth O. Gangel (Victor Books, SP Publications, P.O. Box 1825, Wheaton, IL 60187).

2. Learn more about how to motivate people. Howard Hendricks has an excellent tape on this subject entitled "How to Motivate" (Tape Ministry, Campus Crusade for Christ, Arrowhead Springs, San Bernardino, CA 92414).

The objective of this Introduction Section and of the whole *Leader's Guide*, once again, is to make it possible for you to help people be good managers of themselves. My prayer for you and every member of your group is that God will never let you settle for anything less than His best for your lives.

Session One

INTRODUCTION TO *MANAGING YOURSELF* (Chapter 1)
THE SPIRITUAL PREREQUISITES TO MANAGING
YOURSELF (Chapter 2)

SESSION OBJECTIVES (For Each Group Member):
1. To understand why Christians should manage themselves.
2. To understand the spiritual prerequisites for managing yourself.
3. To apply some of these concepts immediately.

LEADER PREPARATION:
- [ ] Pray for God's wisdom in your preparation.
- [ ] Read and meditate on the session objectives listed above.
- [ ] Read carefully Chapters 1 and 2 of *Managing Yourself* at least once.
- [ ] Read the rest of this session's material several times.
- [ ] Determine what overall customizing of the material you plan to do.
- [ ] Prepare for personal illustrations, examples and group member assignments.
- [ ] Arrange for final needs related to facilities, visuals, refreshments, etc.
- [ ] Pray that special needs will be met in the lives of the group members and that the session objectives will be accomplished.

(See the Introduction section of this *Leader's Guide* for more detail.)

CONDUCTING THE SESSION:

15 min. *(Optional—over and above the 45 minutes provided to cover the content.)*

Get acquainted. This is the first session, and if the people in the group are not well-acquainted with each other, it would be good to encourage them to begin to get acquainted. Consider supplying name tags with the first names written in large letters. Your discussions will be enhanced when individuals are able to call each other by name. You may want to have a time over refreshments before the session begins or have each person introduce himself at the beginning of the session.

1 min. Pray. Ask God for wisdom in seeing the need for personal management and for a close walk with Him.

1 min. Ask how many people in the group can think of something they have really intended to do for some time, but have not gotten around to doing it. Then ask how many do some things very frequently (every day or every week) that they consider a waste of time. Almost all people can think of something in answer to each question, which illustrates why they need to learn how to manage themselves better.

1 min. Read aloud the story of the farmer (*Managing Yourself*, p. 1, par. 1). Ask, "Have you ever had days like this?"

1 min. Read the session objectives, or at least introduce the subject to be covered by reading the titles of the *Managing Yourself* chapters covered by this session.

10 min. Ⓥ Ask and lead a brief discussion on the following questions: (To stimulate discussion you might want to write answers people give on your chalkboard or other visual aid.)

1. What are some reasons why we should seek to manage ourselves better? Some possible answers:
   - To be better stewards of what we have been given.
   - So we can get more done.
   - To make the most of our time.

Ⓥ The use of this symbol indicates an appropriate place to consider using some type of visual aid.

2. What are some Scriptures that suggest we should manage ourselves better? Some possible answers:
   - Matthew 25:14-30
   - Ephesians 5:15,16
   - Galatians 5:23

3. How do you hope to benefit from this group study series? Answers are personal and will vary.

5 min. Paraphrase or read to the group the last paragraph in Chapter 1 and the first paragraph in Chapter 2 of *Managing Yourself* as a transition into the second part of this session: understanding the spiritual prerequisites to managing yourself. If possible, briefly tell the group of your own experience in becoming a Christian and in learning to have a close fellowship with God. Give an illustration of how He has given you wisdom or other help in your daily activities.

21 min. Highlight the content of the four prerequisites to managing yourself found on pages 11-20 of *Managing Yourself*. You can do this by lecturing from your book without their referring to their books or by simply guiding them through pages 11-20 of *Managing Yourself* while they follow in their books. In both cases, be sensitive to the level of spiritual maturity of the individuals in the group.

Cover more thoroughly the concepts you believe to be especially needed and only touch on content you believe to be well understood and applied by people in the group. For example, if you know everyone in the group is assured of the fact he is a Christian, you can spend less time on Prerequisite 1. The following are some questions you may want to ask at the end of your highlighting of each prerequisite. This will help keep the group thinking with you.

1. At the end of Prerequisite 1 ask the group how a person becomes a Christian. Some possible answers:
   - He asks Christ to come into his life.
   - He asks God to forgive his sins and trusts Him to save him.

2. At the end of Prerequisite 2 ask the group in what ways close fellowship with God is broken by sin and what can be done to restore the close fellowship. Some possible answers:
   - Because God doesn't sin and doesn't fellowship closely with those who do.
   - We need to agree with God concerning our sins and turn from them to restore close fellowship.

3. At the end of Prerequisite 3, ask the group where the power comes from to live the Christian life and how it is activated in the life of the Christian. Some possible answers:
   - The power comes from God in the form of His Spirit.
   - Christians can activate this power by trusting God to fill them with His Spirit.

4. At the end of Prerequisite 4, ask the group to list some of the elements involved in a Christian walking in the Spirit. Some possible answers:
   - By confessing sins to God as soon as he becomes aware of them.
   - By continually trusting God to keep him filled with His Spirit.

5. Also after Prerequisite 4, you might want to ask the group what some of the connections are between walking in close fellowship with God and managing yourself. Some possible answers:
   - God will supply us with wisdom moment by moment in how to spend our time, if. we walk closely with Him.
   - God will give us power to follow through on the things we should do if we walk closely with Him.

If you don't have time to ask questions 1 through 4 above, you might just highlight all four prerequisites in the text, then ask the group question 5 above.

3
min. When you are through highlighting, tell the group that in closing this first session you would like to have a time of

prayer. Each individual should pray silently, asking God to reveal any needs he may have in line with the spiritual prerequisites mentioned above. Suggest that he admit his need to experience forgiveness for his sins, asking God to take control of his life, committing himself to a closer walk with God day to day. (Of course each will pray as is appropriate with his own needs in mind.) After the time of silent prayer, you close by praying aloud, thanking God for His answers to the many prayers He has just heard.

2 min. Tell the group that you will make yourself available to discuss the content of this session at greater length with anyone who would care to do so. Encourage each who has made a spiritual commitment to share this with someone close to him or her. This will bring personal encouragement, especially if his or her friend prays for him or her. Indicate the importance of such fellowship to promote spiritual growth. Asign the reading of Chapters 3 and 8 of *Managing Yourself* for the next session. (Also assign the reading of Chapters 1 and 2 if the group members have not read these chapters before this session.) Paraphrase or read the Session 2 objectives to the group to give them some initial motivation to read the two assigned chapters for themselves in anticipation of the next meeting.

ADDITIONAL MATERIAL (Supplementary Content Ideas):

1. Study the various Scriptures mentioned in Chapters 1 and 2 of *Managing Yourself* in greater depth. For example, Matthew 25:14-30. The following are some ways to dig deeper into the passage.

   a. Read the passage as a group and discuss it.

   b. Divide the group into 2 or 3 subgroups and have them each read and discuss the passage and report back to the whole group what they discovered.

   c. Have a few people in the group prepare a skit—to be acted out during the session—which dramatizes the master's initial entrusting of the talents to the servants, the investment and return process of the first and second servants, the burying of the one talent and the final interaction of the master with his servants.

2. Have the group members write down how they spent each 15-minute period in a one- or two-day period, and evaluate what percent of their time is not very productive in their opinion. Discuss their findings in the next group meeting.

3. Consider the following as thought questions, either for group discussion or personal reflections:

    a. Do you sense you often waste time? If so, why do you think that occurs?

    b. If someone were to ask you what you hope to accomplish in life, what would you say? Are you satisfied that your answer is a good one?

    c. Do you keep putting things off that need to be done? What do you think is the solution to that problem?

4. Have each member of the group memorize Ephesians 5:15 and 16.

5. Assign for individual members of the group to read, and if possible discuss with you, specific Transferable Concepts booklets in their particular area of need or question. For example, if a person in your group is not sure whether or not he is a Christian, have him read the booklet "How to Be Sure You Are a Christian" and discuss it with him afterwards. The following is a list of the Transferable Concepts. If they aren't available in your local Christian bookstore, you can order them from: Here's Life Publishers, P.O. Box 1576, San Bernardino, CA 92402.

1. *How to Be Sure You are A Christian*
2. *How to Experience God's Love and Forgiveness*
3. *How to Be Filled With the Spirit*
4. *How to Walk in the Spirit*
5. *How to Witness in the Spirit*

6. *How to Introduce Others to Christ*
7. *How to Help Fulfill the Great Commission*
8. *How to Love By Faith*
9. *How to Pray*

Session Two

PLANNING LONG RANGE (Chapter 3)
MORE ON PLANNING LONG RANGE (Chapter 8)

SESSION OBJECTIVES (For Each Group Member):
1. To understand why and how Christians should plan.
2. To write down life objectives.
3. To select one specific area where accomplishment or improvement is desired in the next six to twelve months.
4. To determine the best way to achieve this accomplishment or improvement.

LEADER PREPARATION:
☐ Pray for God's wisdom in your preparation.
☐ Read and meditate on the session objectives listed above.
☐ Read carefully Chapters 3 and 8 of *Managing Yourself* at least once.
☐ Read the rest of this session's material several times.
☐ Determine what overall customizing of the material you plan to do.
☐ Prepare for personal illustrations, examples and group member assignments.

☐ Arrange for final needs related to facilities, visuals, refreshments, etc.

☐ Pray that special needs will be met in the lives of the group members and that the session objectives will be accomplished.

(See the Introduction section of this *Leader's Guide* for more detail.)

CONDUCTING THE SESSION:

1 min. Start with prayer. Ask God for wisdom for the group members in their planning.

3 min Ask if some had specific opportunity to apply some of the concepts learned last week in their daily lives. Have one or two group members briefly give the details. Use their comments as a springboard for you to give a one-minute review of what was covered in the last session.

4 min Read the session objectives, or at least introduce the subject to be covered by reading the titles of the *Managing Yourself* chapters covered by this session.

Next, paraphrase or read the first three paragraphs in Chapter 3 of *Managing Yourself*, pages 23, 24. (You could instead share a personal example here of how you discovered it was important to plan.) Then ask the group what they think are some reasons it is important to plan ahead. Some possible answers include:

• I do a better job when I plan.

• It is less frustrating to those around me when I plan.

• Christ advocated thinking ahead (Luke 14:28-32).

(To further stimulate discussion you might want to use your chalkboard or other visual aid to jot down group members' ideas as they give them.)

2 min. While you are still at the chalkboard, refresh them on what they have read on how to plan by reading aloud or paraphrasing the two paragraphs listed under the heading "How to Plan" on page 26 of *Managing Yourself.* As you talk, write the outline on the chalkboard:

1. Pray
2. Establish objectives
3. Program
4. Schedule
5. Budget

As indicated in the second paragraph, explain to the group that they are now going to have a chance to apply this to their lives.

(From here you basically are going to lead them to implement what is asked of them on pages 26-29 of *Managing Yourself*. If you have another 30-45 minutes over and above the 45 minutes of the suggested time schedule, you might choose to do the same planning in a more elaborate, thorough way by guiding them through pages 92-106 of *Managing Yourself*. Whichever you decide to do, you should be very familiar with the "pages 92-106" passage since it provides added, helpful background information. The following assumes you will use pages 26-29 and provides you with some guidelines in addition to those in the book.)

3 min. Read or paraphrase the two paragraphs under the heading "First Pray" on page 26 of *Managing Yourself*. Then have a time of prayer and encourage each person to silently ask God for wisdom for planning for his life. You close by praying aloud.

13 min. ⓥ Read or paraphrase the four paragraphs under the heading "Determine Your Life Objectives" on page 27 of *Managing Yourself*. Emphasize that there are objectives we all have in common as Christians, as indicated in paragraphs two and three. (This is explained further on pages 93-96.)

But also stress that God intends for each of us to play some specific role in life, as indicated in paragraph four. (This is explained further on page 97.) Very common, specific plans are to be a husband and father or wife and mother and perhaps to have a certain vocation. As we do these, our challenge is to glorify God in them. It is not sufficient, for example, for a man to have as one of his life objectives simply *to be* a husband and father. He should set as his objective to be a husband and father *in the manner prescribed by the Scriptures*, which would include his being a disciple and ministering to (discipling) his family.

Encourage them not to be afraid to dream about what they really hope to see happen. For example, if there are mothers in your group, encourage them to write down their objectives for their children. Ask them to think particularly about what kind of people they would like to see their children become, versus just what career they might pursue. The objectives we have in common as Christians focus on our relationship with God and our conformity to the character and activities of Christ.

After your initial instructions and explanations, allow the group members to work on their life objectives for about seven to eight minutes. There should be no extraneous conversation going on in the group during this time. (A possible exception is for husbands and wives to interact, although it is often best to let them plan separately during this session and compare notes later.)

8 min. When you break the silence after the planning time, explain to the group that they should do more thinking on their life objectives as part of their assignment for next week. Congratulate them on having done as much as they have in writing. Just writing down life objectives greatly increases the chances that they will be accomplished. However, explain to them that they must also have vision and excitement about what they hope, pray and plan to be and to do. Proverbs 29:18 tells us, "Where there is no vision, the people perish" (KJV). In Ephesians 1:15-2:10, Paul prays that "the eyes of your heart may be enlightened" and that "you may know . . . what is the surpassing greatness of His power toward us who believe." The conclusion of this passage (Ephesians 2:10) is that we should walk in the good works that God has specifically prepared for us.

In other words God really *does* have a plan for our lives as Christians to be implemented through His power in our lives. We should be excited to find out what it is and motivated to implement it. Encourage your group members to pray for themselves and for one another, that they will gain God's vision for their lives and that they will not just think of small, ordinary-human-sized objectives, but rather of large, stretching, God-sized objectives for their lives.

Next, read or paraphrase the one paragraph under the heading "Determine a Shorter-term Objective" on pages 27 and 28 of *Managing Yourself*. (There is more background for you on this on pages 98-102.)

The rationale for having them pick only one shorter term objective to emphasize in the next six to twelve months is simply that a lot of mental, emotional and physical energy is often consumed in seeing change in our lives. Many people can effectively handle special achievement or improvement in only one area at a time. Most would give up if faced with 10 such areas all at once. If each truly picks the areas of accomplishment or improvement that is the most significant to him or her, there is no better, simpler plan he or she can be implementing. (See page 29 in *Managing Yourself* for more insight on this point.) Examples of such shorter-term objectives would be:

- Improve my personal devotions to the point of having a regular, quality time each morning.
- Get in shape through an exercise program.
- Lose 20 pounds.
- Finish the fence in the back yard.
- Find a good, regular means of being active in my church.

After your instructions and explanations, give the group three to four minutes to determine their number one, six-to-twelve-month objectives.

8 min Next, read or paraphrase the first four paragraphs under the heading "Determine the Best Program" on pages 28

Ⓥ and 29 of *Managing Yourself*. (Pages 102-104 provide you with more background on this point.) One of the simplest ways to program is to brainstorm for awhile to create a list of activities which are alternative ways to try to accomplish your objective and then select the best way. Preferably this would be one activity that could totally accomplish your objective, e.g., a diet to lose 20 pounds. In some cases, though, there is a logical first step that will not by itself accomplish the objective. An example would be needing to gather more information as a first step—by reading, asking, etc. Then the person can pick one best alternative. In such a case, he should put the number "1" in front of "Gather more facts" and then anticipate that he will soon have completed that activity and can scratch it from the list, add some new ideas and then pick a number one priority activity that will take him the greatest distance toward the accomplishment of his objective.

After your instructions and explanation, allow the group four or five minutes to determine their number one priority activities toward their number one objectives for the next six to twelve months.

3 min. In the course of this session the group members have accumulated some notes related to personal applications of the concepts covered. For example they should have rough drafts of their respective life objectives. They should each keep a planning/scheduling notebook or file where they can place these notes after each session. Encourage each group member to start such a notebook or file with the notes from this session. This should in no way replace completing their 3 x 5 cards as indicated below.

Next read or paraphrase the next-to-last paragraph on page 29 of *Managing Yourself*. Give them the assignment for next week of completing their 3 x 5 cards and bringing them to the group for the next session. Also assign them to read Chapters 4 and 10 in *Managing Yourself* before the next session. Close by reading or paraphrasing the last paragraph on page 29 and praying.

ADDITIONAL MATERIAL (Supplementary Content Ideas):

1. As mentioned in the "Conducting the Session" section, you may want to cover this whole session more elaborately, taking more time. If so, follow the instructions on pages 92-106 of *Managing Yourself* as your guide.

2. At various points in the "Conducting the Session" flow of content, discussion could be used to great advantage as time allows. After each time you ask the group to write on their own, you can start a discussion by simply asking if there are any questions or by asking if anyone in the group would like to share with the group what he or she has written. A little time will be spent in this kind of discussion at the beginning of the next session.

3. As a homework assignment, ask the group to go back through their thinking related to the next six to twelve months. Using pages 98-106 as a guide, have them think more thoroughly about accomplishing their number one priority objectives. Have them submit their plans to you in the format of the example on page 106.

4. Study more thoroughly some of the Scriptures mentioned in Chapters 3 and 8.

   Particularly appropriate are the passages mentioned on pages 94 and 95 of *Managing Yourself*. The following are some ways to do this:

   a. Simply read the passages to the group and discuss them.

   b. Divide the group into subgroups and give each subgroup a different passage to read and discuss. Have them report back to the whole group on what they find.

   c. Have someone in the group do a dramatic reading of Revelation 4 and then Psalm 96. Have people comment on how they felt as the passages were being read and how that contributes to their understanding of the objective of glorifying God.

5. Have each member of the group memorize I Corinthians 9:24-26.

Session Three

SCHEDULING YOUR TIME (Chapter 4)
SCHEDULING YOUR WEEK (Chapter 10)

SESSION OBJECTIVES (For Each Group Member):

1. To understand why Christians should schedule their time.
2. To learn how to select and schedule the priority activities from all the possible things to do.
3. To learn to apply these concepts to a brief period of time.
4. To learn to adapt these concepts to apply to scheduling a day.

LEADER PREPARATION:

- ☐ Pray for God's wisdom in your preparation.
- ☐ Read and meditate on the session objectives listed above.
- ☐ Read carefully Chapters 4 and 10 of *Managing Yourself* at least once.
- ☐ Read the rest of this session's material several times.
- ☐ Determine what overall customizing of the material you plan to do.
- ☐ Prepare for personal illustrations, examples and group member assignments.
- ☐ Arrange for final needs related to facilities, visuals, refreshments, etc.

☐ Pray that special needs will be met in the lives of the group members and that the session objectives will be accomplished.

(See the Introduction section of this *Leader's Guide* for more detail.)

CONDUCTING THE SESSION:

1 min. Start with prayer. Ask God to use this session to help the whole group recognize what are the priorities in their lives and be able to plan to spend their time on those priorities.

5 min. Ask if some of the group would like to tell the rest of the members what kinds of things they have written down on the 3 x 5 cards which they were to bring to this group meeting. Recognize that some of the planning that they had done is rightly very personal and private. On the other hand, a number of things that people are intending to do would be very appropriate to share with the group. Write some of these down in your notebook and encourage the rest of the group to do the same, so that they can later pray that God will be doing these things in the people's lives. At the end of this sharing time, take about 30 seconds to one minute to review what was covered last session.

5 min. Read the session objectives, or at least introduce the subject to be covered by reading the titles of the *Managing Yourself* chapters covered by this session.

Ⓥ Ask the group to answer the questions, "What are some reasons that Christians should schedule their time?" and "What are some Scriptures which support these reasons?" As you receive answers from the group, use your chalk board or other visual aid to note remarks and to stimulate further discussion. You might also use this as an opportunity to organize into small subgroups—with even as few as two people per subgroup—to discuss the question first. Afterwards have people share their ideas with the whole group. Some possible answers to this question include:

- If they don't, they will be wasting their time and thus not truly glorifying God.

- The Scriptures indicate that we should, "Make the most of your time" (Ephesians 5:15,16).
- If we don't, we won't get as much done, as happened to the Israelites in Haggai 1.

You will find other background and answers to this question on pages 31-34 of *Managing Yourself.*

10 min. Ⓥ While you are still in front of the visual aid, highlight the four steps of how to schedule your time as found on pages 35-41 of *Managing Yourself.* Include highlights also of the example found on these pages. As you are doing this, list on the visual aid the four main points:

1. List activities.
2. Ask if assignable.
3. Assess priorities.
4. Schedule.

As with previous highlighting, you can either lecture from your book with group members having their books closed, or you can have them leaf through their books with you as you speak. Don't feel the need to elaborate extensively on what is in the book, but simply highlight the main points. This will insure that all members are familiar with the material, though they may not previously have read or understood it.

10 min. Have the group apply the "how to schedule your time" outline to the next available two-to-three-hour period of time in which they would normally be awake. What they are going to come up with will look very much like the example detailed on pages 39-41 of the text. In particular, their list should look something like the list shown at the bottom of page 40 with names and priorities to the left and with time slots written to the right of some of the activities. In the example in the text the number one priority activity, "Review Talk," was scheduled for 1½ hours. If the starting time of the period of time was 3 o'clock, then the time slot that could be put to the right of that item on the list is "3-4:30." Alternatively, they may prefer to enter the first few of their priority items in the schedule/calendar they use day to day, putting their number one priority first in the time period being scheduled.

Normally it is helpful to get the group started by saying, "Now let's begin by listing activities," then encouraging them to move on to asking if it is assignable and assessing priorities for the next couple of minutes. Often it is helpful to walk around the room and encourage people individually to keep moving on the assignment.

9 min. ⓥ Highlight for the group the content under the heading "Some Refinements in Your Scheduling" found on pages 41 through 44 of *Managing Yourself*. You can handle this much as you have the other highlighting. Or, in this case, you might choose to assign each of the seven points made in this section to one or two members of the group to work on as individuals or as pairs. Give them a couple of minutes to prepare and then have each point reported on for 30 to 45 seconds in quick succession.

If in your group you have mothers of young children or other people who do not have much say over how they spend their time or who are frequently interrupted, particularly emphasize the first paragraph of the first refinement (on pages 41 and 42 of *Managing Yourself*). I doubt if God will ever ask a young mother why she did not do some elaborate project while the baby was screaming and the pot was boiling over. In those circumstances her top priorities are obvious. But I do believe God will hold even young mothers responsible to make the most of their time (however little is discretionary) where they are able to make progress toward their number one priority activities. Everyone should at least have a list, know what the number one priority activity is, and look for time in which to do it.

Some homemakers have told me that they really benefit from having a routine schedule for at least the first part of the day. After sending the husband to work and some of the children to school, they like to focus on getting dressed, doing the dishes, and making the beds before tackling anything else. Otherwise the "mess" causes them to feel disorganized and inefficient. By all means, the determination of standard schedules and priorities is subject to how each person works best.

Some homemakers (and other people for that matter) say they like to have a main theme for each day such as: Monday—washing and cleaning the house, Thursday—shopping, Saturday—taking the children to their activities.

5
min. Ask the group members to try to schedule for at least one full-day period before the next session (assuming the sessions are at least that far apart). They mainly will be applying the four-point outline of listing activities, determining if they are assignable, assessing priorities and scheduling. However, the refinements that have just been highlighted will be very valuable to them as they consider scheduling a whole day. Even in these few minutes it would be good for them to see the application of some of the refinements to their daily lives. One important aspect of scheduling a day is deciding which normal activities are to be considered as standard and not worthy of being listed and prioritized, and which activities are those over which we should exercise daily discretion. Encourage them to take a minute to begin to determine the standard activities in their days.

Indicate that they should bring at least one day's schedule to the next session, along with any questions they have concerning it. Have them refer to pages 44-47 in the text for specific help in scheduling a day. Encourage them to make a habit of scheduling either for brief periods or for entire days. One or the other should be practical for most of the group. Also mention to the group that each person should have some sort of schedule/calendar on which he can record future commitments. (See point 1.e. on page 71 of *Managing Yourself.*) Show them what kind you recommend using.

Encourage the group members to place their personal

application notes from this session in their planning/ scheduling notebooks or files.

Assign Chapter 5 and 6 of *Managing Yourself* for the next session, then close in prayer.

ADDITIONAL MATERIAL (Supplementary Content Ideas):

1. Consider covering the content of Chapter 10 in some detail with a group who could benefit from that level of scheduling. Most people, I have found, have all they can do to schedule two to three hours at a time, or perhaps one day at a time. If, however, the people you are working with have very complex activities and need to be committing to things well in advance, it would be helpful to teach them how to schedule for a week. As you can see from looking at the chapter, it is a good deal more complicated, but is well worth the effort under certain circumstances.

2. After the group tries out the four-point outline on their own two-to-three-hour time periods, ask them if they have any questions. You might also ask if they discovered that they would now be doing certain activities that they would not otherwise have done had they not scheduled. Some people may suggest that otherwise they would have done a few little odds and ends to get them out of the way. At this point you can point out that such a practice often can prohibit the truly priority things from ever getting done.

3. Study in more depth the various Scriptures mentioned in Chapter 4 (e.g., Haggai 1, Matthew 6:33, II Thessalonians 3:1, I Thessalonians 5:21). A particularly good Scripture to study is Haggai 1:1-11. You might have one person write a modern paraphrase of that passage (e.g., as if Haggai were writing to today's businessman or factory worker or homemaker) in preparation for the session. Have him read it to the group and hand out copies of it. Then divide the group into "breakdown groups" to discuss it. Have these groups report on their discussions.

4. Consider the following as thought questions, either for group discussion or personal reflection.

  a. What are some meanings of the word "priority"? Is it the same as urgency? If not, what are some reasons it is not? Will each individual perform the same activities as everyone else if they do what is priority? If not, what are some reasons the activities are not the same?

  b. What are some practical limits on what can be scheduled in advance? In what ways does the existence of these limits not cancel the need to schedule what can be scheduled?

5. Have each member of the group memorize I Thessalonians 5:21.

Session Four

FOLLOWING YOUR SCHEDULE (Chapter 5)
MULTIPLYING YOUR TIME (Chapter 6)

SESSION OBJECTIVES (For Each Group Member):

1. To understand how to follow through on what is planned and yet stay relaxed and sensitive.
2. To apply these principles in some very practical situations.
3. To understand "why's" and "how to's" for doing the best job in each activity that is done.
4. To apply a few of the ideas related to objective 3 right away.

LEADER PREPARATION:

☐ Pray for God's wisdom in your preparation.
☐ Read and meditate on the session objectives listed above.
☐ Read carefully Chapters 5 and 6 of *Managing Yourself* at least once.
☐ Read the rest of this session's material several times.
☐ Determine what overall customizing of the material you plan to do. This session can easily go more than 45 minutes because of the number of discussion questions. If your group tends to discuss each question for a long time and you are limited to 45 minutes, you will probably want to delete some of the content. For example,

you might pick just two of the scenarios with Fred which seem to be in the areas of greatest need in your group. Also, you might delete one of the three scheduled content ideas with regard to Chapter 6.

☐ Prepare for personal illustrations, examples and group member assignments.

☐ Arrange for final needs related to facilities, visuals, refreshments, etc.

☐ Pray that special needs will be met in the lives of the group members and that the session objectives will be accomplished.

(See the Introduction section of this *Leader's Guide* for more detail.)

CONDUCTING THE SESSION:

1 min. Start with prayer. Pray for the members of your group, that they will learn the balance between following through on what they set out to do and being sensitive and flexible to new situations that they might face. Also, pray that they will learn to be even better stewards of their time in learning how to do each activity in the best possible way.

3 min. Ask the group how they were helped by applying the scheduling principles and techniques learned at the last session. They were to bring one day's schedule to the group session this time. Use those comments as opportunities to highlight what was covered last week. If anyone mentions difficulties in actually following through, indicate that "following through" is what you are going to cover today. If you can take longer than 45 minutes on this session, ask if they have some questions now. Otherwise encourage them to ask questions after the session.

The following four items in the flow of content are similar. In each case you will lead off by reading a brief scenario depicting a problem that a person (a Christian) is facing, then ask the group how they would recommend the person solve his problem. In sequence, the emphasis of each case will be the points made in Chapter 5 on how to follow the schedule: motivation, discipline, sensitivity and peace. As you conduct these discussions, try to lead the group toward the main point each scenario poses.

Try to cause them to see how that point from the text (motivation, for example) could be made to apply to their particular situation. Your chalkboard (or other visual) will help focus the discussion on the analysis of each of the four situations.

6 min. Read the session objectives, or at least introduce the subject to be covered by reading the titles of the *Managing Yourself* chapters covered by this session.

Explain to the group that you will be giving them some specific situations. Indicate that you are going to ask them how they would help the person in the situation in light of what they have been learning in Chapter 5 of *Managing Yourself.* The following is the first story to read to them:

> Fred had been at his garage workbench for about 15 minutes trying to get started on a project he had promised his wife he would do today. As he looked around the workbench, he saw a number of other projects that he also needed to do, all of which appealed to him a great deal more than this particular project. Many of them were even similar to the one he was contemplating.

Ⓥ After you read the story to the group, ask them what advice they would give Fred. This story tends to lend itself to two of the points in Chapter 5. One is discipline and the other is motivation. If possible, see if you can feature the motivational aspects of the situation.

First, since Fred is doing this project for his wife, his love for her should motivate him to complete the job. (Sadly, sometimes the love that would motivate a husband is lacking.)

Second, since Fred obviously enjoys doing projects such as this one, if he were to begin working on the aspect of this one that he enjoys, it could spark the necessary motivation to complete the entire project.

Third, according to Philippians 2:13, it is God who works in us to give us desire and power to do things that please Him. It pleases God for Fred to show love for his wife. Therefore, Fred should ask God to provide him with the necessary motivation.

5 min. Read the following story to the group and conduct a discussion in a similar fashion to the point above:

> For a number of weeks now, Fred has been meaning to get started on an assignment which his boss gave him at work. He basically likes to do this kind of assignment, but it seems that different things have been coming up that keep him from getting it done. His boss has already criticized him once in light of how late the work has become.

Ⓥ Ask the group what advice they would give Fred. Here again, there are certain motivational aspects, but the point for you to feature in this story is the need for discipline. Fred basically likes to do this kind of activity, but when confronted with several opportunities to follow through on it, he has procrastinated. The group should be loaded with similar illustrations from their own experience from which they can offer practical advice on how to make discipline a part of Fred's life. Finally, indicate that discipline is available from God as a part of the fruit of the Spirit (Galatians 5:23).

5 min. Read the following story to the group and conduct a discussion in a similar fashion to the two points above:

> During work today, Fred agreed to go bowling with some of his fellow workers. When he got home for supper, he found out that he had forgotten that he had promised to accompany his son to the boy's little league baseball game. When he mentioned that he had a commitment to go bowling that evening, he noticed a tear forming in the corner of his son's eye.

Ⓥ Ask the group what advice they would give Fred in this case. Obviously the key issue here is one of being sensitive to people. Actually, there are now two sides of the sensitivity question. Fred is in the position of having committed himself to several different people, so he really needs to pray for God's wisdom.

Most people would agree that Fred's first responsibility is to take his son to the ballgame. However, when he does this, he will have to explain the situation—and his decision—to his fellow workers in a sensitive way. You can use this story to draw from the group a number of practical tips they will have regarding being sensitive to situations such as this one.

5 min. Read the following story to the group and discuss it in a manner similar to the three previous stories:

> Fred had a terrible day at work. It seemed like everything went wrong. He was misunderstood in his intentions when he offered to help on someone else's project. When he got home, he learned that his son had gotten into trouble at school and had been disciplined by the teacher. He decided to watch some T.V., but turned it on only to discover that it needed repair. Fred was really frustrated.

Ⓥ Ask the group what advice they would give Fred at this point. Clearly the issue is one of learning to experience God's peace. There might be some things that can be done to unravel some of the different situations, but despite all that, Fred needs to learn to experience peace. It would be good for the group to really focus on how God provides that peace in spite of circumstances. Also, pull out of the group various practical pieces of advice on how one can begin to experience God's peace, such as by "counting your blessings," etc.

5 min. When finished with the above discussion, announce that you are moving to the content of Chapter 6 in the text. Have them turn in their Bibles to Colossians 4:2-6, which discusses a quality-conscious way of life. Have someone read it to the group, then discuss how the verses illustrate God's intention that we do things well. For instance,

instead of just praying, we are to "devote ourselves to prayer." We are to keep ourselves very alert. We are to possess acceptable attitudes; to speak clearly; conduct ourselves wisely, etc. This is a good Scripture to discuss first in groups of two or three, then summarize in the whole group.

6 min. Next, tell the group that they are going to be applying two of the five points mentioned on pages 66-69 in *Managing Yourself*. Have them turn to "Establish Objectives" on page 66. Think of a project or activity with which your group would all be fairly familiar and that has some room for improvement in efficiency. Be very careful not to pick something that would be an object of undue negative criticism or could be hurtful to anyone. This activity could be a meeting that has become more traditional than functional and interesting.

Ⓥ Pick such an activity and ask some of the questions listed under the "Establish Objectives" heading. See how the group responds. The idea is to see if you can quickly establish precise objectives that need to be accomplished. In light of these objectives, ask several questions from the first two paragraphs under "Program" on pages 67 and 68.

Keep an eye on your time as you do this, since you are not seeking a total solution, but are basically trying to determine: (1) Does the activity have an objective? If yes, what is it? (2) If it does have objectives, are the various related program activities contributing effectively to the objectives? And, (3), are there other activities that could contribute better?

For example, some meetings are called simply to keep certain constituents informed. Such objectives often can be attained more efficiently by sending out a regular newsletter in lieu of the meeting.

8 min. Have the members of the group quickly scan through the various items listed under the heading "Some Specific Helpful Hints" on pages 71-76 in *Managing Yourself*. Ask them to find one that seems particularly applicable to them in an area where they could improve, and be prepared to mention it to the group. Give them a few minutes to do this and then quickly ask each person to report what he selected and why. Have group members take notes so they can pray for people during the course of the next week. Announce that at the next session you will provide a brief time for individuals to report on how well they did in improving in this particular area.

1
min.
In addition to the assignment just mentioned, ask the group to skim over Chapter 7 and to read more thoroughly Chapter 9 for next week. Encourage the group members to place their personal application notes from this session in their planning/scheduling notebooks or files. Then close the group in prayer.

ADDITIONAL MATERIAL (Supplementary Content Ideas):

1. Study some of the Scriptures mentioned in Chapters 5 and 6 that have not already been included in the session. One very good passage to study is Matthew 14:13-23. One of the great questions that arises as a person walks through life is that of how to balance tasks to be accomplished with the flexibility needed in light of people's needs. Here Jesus confronts that very dilemma. He starts off to do something, is deterred from that by the multitude, meets the needs of the multitude, but then finally accomplishes what He originally intended to do.

   Another passage that illustrates this same point in a little different way is Mark 1:21-45. Here Jesus clearly establishes His priority of teaching and preaching the message that the Father has given Him. However, when the people responded more to His healing ministry, He, out of compassion, responded and met that need. In verses 38 and 39, though, we see His continued emphasis on accomplishing what He set out to accomplish.

   These passages can be assigned as self-study projects or can be discussed as a group. They can even be acted out in a skit and then discussed.

2. Assign one of your group members to read one of the books mentioned on page 223 on the subject of time. Ask him to report how the book relates to the subject of scheduling one's time.

3. Have someone bring a complete log of his activities covering several days. Ask him to give a brief report concerning how he spent his time and the kinds of problems encountered. Then let the group brainstorm various solutions to the problems he faced. An example of this is found on pages 59-61 in the book.

4. In similar fashion to that outlined in "conducting the Session" above, select another activity that is well known to the group. Asking God to give them wisdom, have the group determine how well they think the activity is meeting its stated objectives. Have the group analyze the activity to: (1) Determine if it could be streamlined. (2) Determine what new program activities might more effectively replace some of the older, less effective ones. (3) Select the top priority activities for implementation.

This exercise could be especially meaningful if a "real-life" activity could be found for which the group could be assigned the actual responsibility of doing the above. Thus energies could be directed to solving a problem instead of serving merely as an illustration.

5. Have the members of the group memorize II Corinthians 8:10-11 and/or I Corinthians 10:31.

Session Five

KNOWING GOD'S WILL FOR YOUR LIFE (Chapter 9)
REVIEW

SESSION OBJECTIVES (For Each Group Member):

1. To review the content that has previously been covered.
2. To understand nine principles involved in discerning God's will.
3. To apply these principles in a specific, current situation.

LEADER PREPARATION:

- ☐ Pray for God's wisdom in your preparation.
- ☐ Read and meditate on the session objectives listed above.
- ☐ Read carefully Chapters 7 and 9 of *Managing Yourself* at least once.
- ☐ Read the rest of this session's material several times.
- ☐ Determine what overall customizing of the material you plan to do.
- ☐ Prepare for personal illustrations, examples and group member assignments.
- ☐ Arrange for final needs related to facilities, visuals, refreshments, etc.
- ☐ Pray that special needs will be met in the lives of the group members and that the session objectives will be accomplished.

(See the Introduction section of this Leader's Guide for more detail.)

CONDUCTING THE SESSION:

1 min. Start with prayer. Pray that the members of the group will be given a good perspective in their review of what they have learned and will be helped in understanding by the review. Also, pray that God will give them special wisdom as they consider knowing His will for particular situations that they face.

3 min. In the last session each member of the group selected one particular helpful hint out of pages 71-76 in *Managing Yourself.* Ask the group in what ways they were successful in implementing the particular improvements they were seeking this past week. You might also ask, if time allows, how they did in following their schedules better in light of the four principles learned in Chapter 5. As people share, seek opportunity to highlight some of the points made last week.

15 min. Read the session objectives, or at least introduce the subject to be covered by reading the titles of the *Managing Yourself* chapters covered by this session (Chapters 7 and 9).

Ⓥ Announce a review of the concepts they have previously learned in this study. On a previously prepared visual aid show the group the five chapter titles of Chapters 2 through 6 of *Managing Yourself.* Taking each title in order, ask the questions: "What are some ways in which the content in this chapter is important to our managing of ourselves?" And, "What are some ways in which you have particularly benefited from the chapter's concepts?" The review on pages 79 and 80 of *Managing Yourself* will give you some background for this discussion. Also, if you feel the need, you could review the first part of each chapter where the concepts are defined. In summary, though, the answers to the "why important" question in each of the five may include the following answers:

- *Spiritual Prerequisites*—God is the source of all wisdom. God gives power to implement what is best for us in our lives. God is available to us for moment-by-moment access.

- *Planning Long Range*—We must know where the finish line is in order to run the race in the right direction. We and those around us will be more effective if we think ahead in some detail. God has promised to give us direction (Psalm 32:8).

- *Scheduling Your Time*—Generally speaking, things don't get done unless we allocate some time to them. It is important that we know what is priority and stick to it, not only for our own effectiveness but also for God's blessing on our activities (Haggai 1). Time is a limited commodity which deserves to be spent carefully.

- *Following Your Schedule*—Often we won't follow through with what we intend to do unless we become motivated and/or disciplined. As we are confronted by situations, we need to make many decisions concerning what God wants us to do and what is most loving toward people. It is easy to become frustrated by the day-to-day activities if we don't learn to experience God's peace in following our schedules.

- *Multiplying Your Time*—The Bible requires us to be good stewards of what we are given (Matthew 25:14-30). We are to do things in a manner that is most glorifying to God (I Corinthians 10:31). We often drift into doing things in an inefficient manner if we don't frequently ask if we have clear objectives for the activity and are straightforwardly pursuing those objectives.

2 min. Indicate to the group that they are now beginning the study of Chapter 9, "Knowing God's Will for Your Life." In order to do this best, explain you would like to have each person take a circumstance he or she is now facing, and—by applying the principles stated in this chapter—

work through his or her situation to a solution. Have them try to select situations that can be shared with the group at the end of this session and are not too complex. Before going on, give the group a minute or so to think of the situations they will plan to deal with.

.20 min. (V) Before class, record the nine main points of Chapter 9 on your visual (page 109 through page 114). Use this visual during this time. To implement the first two points have the group spend a couple of minutes in silent prayer, during which time they can check their relationships with God, be sure they are filled with the Spirit and ask God for wisdom in their particular situations. At the end of the prescribed time, you should lead in closing prayer.

Next have the group members draw a chart on their note papers as indicated below:

ALTERNATIVES

Pros:					
Cons:					

Have them enter along the top what alternatives they are considering. (If they are just considering a "yes" or "no" on one alternative they need to fill in only one of the top spaces.) As you proceed through the "knowing God's will" outline, they will be entering "pros" and "cons."

Next have the group consider the third point. Highlight or have them read the two paragraphs under "Consider Scriptural Objectives" found on page 110 of the text. Ask them if this point causes them to think of "pros" or "cons" relative to their decision. If so, have them make appropriate entries in their chart. For example, a group member may be considering whether to participate in an additional Bible study one evening per week. From the point of view of scriptural objectives, one "pro" could be "contributes to my being a disciple" and one "con" could be "takes one evening per week of my time away from my children." Some alternatives very obviously do not significantly glorify God and will probably be eliminated by that major "con."

Next move on to "Take All Scripture into Account." Highlight or have them read over that point and apply it to their particular situations. It might be that God has been reminding them of particular Scriptures that seem to give specific wisdom on their situations. In some cases the Scriptures expressly prohibit or encourage certain aspects of what they are considering.

"Consider Facts about Yourself" is a category that particularly relates to determining God's will for our careers or for our entire lives. Any situation can be analyzed better if the facts are out in the open. Have them pause and reflect on their strengths and weaknesses, on their past experiences and existing future plans, and on what they enjoy. Do these facts seem to suggest a pattern of God's leading relevant to this particular situation? If so, have them record the appropriate "pros" and "cons" on their charts.

Continue on through the rest of the points in a similar fashion. With each point ask, "Do I see anything in this particular category that seems to bear on my decision at hand?" In some cases, there will be no particular item that seems to come to mind that relates to that category. They can feel free to move on to the next category in that case. For example, they may not have received any counsel from anyone, although it might be good to seek some counsel if that seems appropriate.

When you come to the point, "Weigh the Pros and Cons," have them look over their charts. In light of the strength and relevance of the "pros" and "cons" and in light of how many there are of each, most often the right alternative becomes apparent. Ask each of them to come to a tentative conclusion.

Highlight or have them read the material in the text under the last point, "Ask for God's Confirming Peace." Lead the group in prayer, asking God for His confirming peace with regard to each group member's decision (especially in matters of great significance and permanence) and thank Him for the time together in this session.

3 min. Ask the members of the group to share what they have decided concerning their particular situations and what they have learned about discerning God's will.

1 min. Encourage them to follow through and implement the decisions they have reached, if they have finally made decisions. Encourage the group members to place their personal application notes from this session in their planning/scheduling notebooks or files. Also, have them read for next week the "Introduction to Chapters 11 through 17" (pages 127 and 128) as well as Chapter 11 of *Managing Yourself*. Assign each of the 11 qualities mentioned in this chapter to a group member to highlight for one minute during the next session. Preferably spread the assignments among the group members as much as possible. In their reports they should mainly describe the quality and, if time allows, mention one or two reasons why it is important.

ADDITIONAL MATERIAL (Supplementary Content Ideas):

1. Assign a member of the group to prepare a five-to-ten-minute summary report on the content covered to date.

2. Ask each member of the group to tell the single most important thing he has learned, and how he has used it.

3. Give the following quote to the group and ask them to respond and refute it: "There is no use in my scheduling. Things always change anyway." Or you may ask two people from the group to play opposite roles in a skit or debate. One would be a person advocating managing himself in light of what he has been learning. The

other would be a person who really feels that "There's no use . . ." as the quote above reflects. After this, have the group discuss it.

4. Study some of the various Scriptures mentioned in Chapter 9. Ask of each one what it adds to our confidence in and understanding of God's revealing His will to us. These Scripture studies could be done as homework by the people in the group, discussed in subgroups, or discussed in the context of the whole group.

5. Ask the group the following questions:

    a. What are some alternatives to seeking God's will?

    b. What are some of the disadvantages of the alternatives?

    The point here is that we would obviously be less effective if we did not seek to know whatever the all-knowing, all-wise God is willing to reveal to us concerning His perfect will.

6. Have the members of the group memorize Psalm 32:8 and/or Psalm 37:23.

Session Six

THE SPIRITUAL AREA OF LIFE (Chapter 11)

SESSION OBJECTIVES (For Each Group Member):

1. To understand more comprehensively how a Christian can increasingly conform to the image of Christ.
2. To select one specific area of needed improvement and plan to implement it.

LEADER PREPARATION:

☐ Pray for God's wisdom in your preparation.

☐ Read and meditate on the session objectives listed above.

☐ Read carefully Chapter 11 of *Managing Yourself* at least once.

☐ Read the rest of this session's material several times.

☐ Determine what overall customizing of the material you plan to do.

☐ Prepare for personal illustrations, examples and group member assignments.

☐ Arrange for final needs related to facilities, visuals, refreshments, etc.

☐ Pray that special needs will be met in the lives of the group members and that the session objectives will be accomplished.

(See the Introduction section of this *Leader's Guide* for more detail.)

CONDUCTING THE SESSION:

1 min. Start with prayer. Pray that God will make this a very meaningful overview of the kinds of qualities of life that should be present in a Christian, and that He will show each person in the group which of these is priority for him at this time and how he should grow in it.

3 min. Ask some members of the group to share how they were able to proceed in the decision areas they worked on last week. Encourage them to share particularly how it was helpful to them to have made a decision and move in that direction.

2 min. Read the session objectives, or at least introduce the subject to be covered by reading the title of the *Managing Yourself* chapter covered by this session.

Tell the group that with this meeting you are beginning seven sessions on different areas of life as presented in the book *Managing Yourself*, Chapters 11 through 17. The introduction to these chapters on pages 127 and 128 can be either read or paraphrased to cover this point. Emphasize to the group that in *each* session they should seek to find *one* very practical area of application for themselves (which may vary widely from member to member). Tell them you realize that it would be very difficult for each person to implement, all at once, the seven different areas of change he or she will discover in the next seven sessions. So what they will really be doing is some preliminary thinking, from which one or more areas of change might be tried over the next several months, as it becomes compatible with other changes that might be going on in their lives.

29 min. Have each member of the group give his one-minute report as assigned in the last session. After each report have the others ask questions and briefly discuss the quality. Proceed in the order in which the character qualities are listed in Chapter 11. The reporter should mainly describe the quality he is highlighting and, if time allows, give one or two reasons why it is important to the Christian. I recommend you appoint a timer to keep each report within limits. As a variation, you might consider giving more time for reporting and allow less time for discussion. As a stimulus to thinking you might ask one or two of the reporters to take a few extra minutes and tell of a few specific ways in which Christians can be encouraged to grow in their areas. These brief reports can be challenging and very good at involving some of the group members in a more significant way.

During each report the people in the group should be making personal application by asking themselves if this is an area that God is laying on their hearts in which they should seek improvement now. It may be of help to the group if you were either to have the 11 qualities listed on your visual aid or to have the group members look at their books as the people present them. In either case, they certainly should be jotting down additional ideas on each quality.

8 min. Have a brief time of prayer in which you ask God to give the people wisdom specifically in choosing their highest priority objectives. Now have each member of the group select the one particular quality that seems to be the highest priority need for him at this time. Have him sharpen exactly what God seems to be telling him. For example, if faith happened to be the area where he sensed the greatest need for improvement, he should ask God if He is suggesting he tackle an already existing situation with greater faith or if he is being directed to seek new challenges beyond those he has previously sought. Or perhaps he is simply being shown how to express to God his true trust in Him with his life.

Once your group members have determined what their objectives are, have them determine what program activities seem to be best. They can use the planning techniques learned earlier.

2 min. Assign the group Chapter 12 to read by the next session. Encourage the group members to place their personal application notes from this session in their planning/ scheduling notebooks or files. Close the session in prayer.

ADDITIONAL MATERIAL (Supplementary Content Ideas):

1. Have one or two members of the group give book reports on some of the books given in the "For Your Further Study" section on pages 136 and 137 in *Managing Yourself*. In preparing their reports, tell them to distill out some of the most important principles mentioned by the authors and highlight them for the group.

2. Present an actual case study of a person who has particular spiritual needs. See if the group can sense which of his spiritual needs are the greatest and then design a plan of action best suited for that person. Normally it would not be wise to deal with one of the group in this fashion. Perhaps you could disguise an example or conjure up a composite example to discuss with the group. Lead the discussion to analyze the character qualities that are most deficient and set up a plan to help correct them.

3. Consider the following as thought questions either for group discussion or personal reflection:
   a. What are some ways in which the spiritual area of a person's life is important?
   b. What would be some characteristics of a person if he did not possess much of any of the qualities listed in Chapter 11?

4. Read a passage from one of the reference books listed or another book of your choice on a spiritual topic and have the group discuss what it has to say about the need for spiritual growth. Instead of a passage suggested above, consider using Ephesians 3:14-21 or some similar Scripture and asking the group: (1) What important points these verses make on growing spiritually, and, (2) What are some ways they suggest a person should grow spiritually?

5. If you have some individuals gifted in drama, you might suggest that they devise a dialogue between two citizens

of the Kingdom of God. For this particular session, the dialogue should portray the Kingdom as though it were an earthly kingdom, with a king, laws, etc. They would be talking about how the laws of the land operate and how the people respond. After they are through with the dialogue, have the group discuss some ways in which the "skit" kingdom differs from the world's kingdom, and some ways in which the character qualities discussed in Chapter 11 would make a person a better citizen of the "skit" kingdom.

6. Have the members of the group memorize Romans 8:29.

Session Seven

THE MENTAL AREA OF LIFE (Chapter 12)

SESSION OBJECTIVES (For Each Group Member):

1. To understand the vast potential of God's gift of the mind and its strategic role in the whole of life.
2. To learn how to tap the potential of the mind better.
3. To select one specific area of needed improvement and plan to implement it.

LEADER PREPARATION:

- ☐ Pray for God's wisdom in your preparation.
- ☐ Read and meditate on the session objectives listed above.
- ☐ Read carefully Chapter 12 of *Managing Yourself* at least once.
- ☐ Read the rest of this session's material several times.
- ☐ Determine what overall customizing of the material you plan to do.
- ☐ Prepare for personal illustrations, examples and group member assignments.
- ☐ Arrange for final needs related to facilities, visuals, refreshments, etc.
- ☐ Pray that special needs will be met in the lives of the group members and that the session objectives will be accomplished.

(See the Introduction section of this *Leader's Guide* for more detail.)

CONDUCTING THE SESSION:

1 min.
Start with prayer. Pray for the members of the group, that they comprehend what a great resource God has given them in the form of their minds and that they will learn specific ways they can use them better.

1 min.
Give a brief review of what was covered in last week's session on the spiritual area of life.

5 min.
Read the session objectives, or at least introduce the subject to be covered by reading the title of the *Managing Yourself* chapter covered by this session.

Ⓥ
Have someone read Proverbs 23:7 and another read Romans 12:2 to the group. Ask the question, "What are some ways the mind is important in our growth as people?" The answers here should center on the fact that it is preeminent in our growth and that our thoughts normally precede our actions. Furthermore, God uses the renewing of the mind to bring about change in our total beings (Romans 12:2). Next have someone read Psalm 19:7,8. Then ask the question, "What are some things this passage tells us about the impact of the Word of God on our minds?" The answer here centers around the fact that the Word enlightens our minds and makes us wiser and more joyful. Even those who would by the world's standards be considered "simple" can be made wise through the Word of God.

13 min.
As indicated on page 140, first paragraph under the heading, "Mental Functions and God's Guidance," a simplified view of the mental functions involves three kinds of activity: receiving, processing and sending. Highlight this for the group and then indicate that you will first be thinking together with them about the area of "receiving." Begin by showing them that it is possible to learn to observe more than they would normally observe. Place a couple of new items in the room where the group has been meeting for several weeks and ask who can tell which are new and which are not. The main point here is to show them that there is a lot of information they are "receiving" which

they may or may not be utilizing, and that the area of observation can be improved.

Next, interview one person in the group, asking him what he did the previous day. As he tells you, keep asking for more detail in a sincere and interested fashion. After you have done this for awhile, explain to the group how this illustrates the possibility of being aggressive and inquisitive in the seeking out of more information and how listening aggressively can have some interpersonal benefits as well as providing more details for our minds. The main point of these exercises is to encourage them to consider sharpening their abilities to receive information. Tell them that "aggressive" interviewing is only representative of the many ways in which we can "receive" more effectively and efficiently. Speed reading would be another illustration.

14 min. **Ⓥ** Mention to the group that now you will be talking about the area of processing information in the mind. Give them an overview of the several different kinds of activities the book mentions that go on within the mind, all of which could be categorized as processing. Take a minute or two to highlight the contents and subheads on pages 145-150.

Next, demonstrate the kind of memorizing they could do if they actually put their minds to it.

Ask them how many could name all 12 of Jesus' disciples quickly. No doubt most, if not all the group, could not do that. Next, read to them the following story which is an association method of memorizing the names of the 12 disciples. Then see how many of them can, by recalling the illustrations, name the disciples. (Dramatization, action, humor, exaggeration, visualization and association are extremely helpful in memorizing a list; the following exemplifies these techniques.)

> Picture a tiny fish thrashing at the surface of a lake with a huge net dragging it into a nearby boat. As you get closer to the boat, you see a strong right arm drawing the net toward the boat. The barechested man to whom that arm belongs represents Simon

Peter, a strong and skillful fisherman. With his left arm Peter is fighting off a net being thrown over him by another man in the boat. This man is Andrew, who brought Peter to Jesus or, if you will, netted him for Jesus. (Be sure to dramatize the action of Peter with an overly strong right arm bringing in a tiny little fish and then with his other arm fighting away this nuisance of a net being thrown over his head by Andrew.)

Andrew is standing up in the boat, yet is unable to stand still because bolts of lightining coming out of two thunderclouds above are striking the boat, sending geysers of water spurting up around him. The two thunderclouds represent the two Sons of Thunder, James and John.

At the same time, rain is pouring from the clouds directly into a huge barrel that is being filled up to overflowing. The filled-up barrel represents Philip.

Next picture all the water that is overflowing from this huge barrel splashing onto a dog and a cat that are tied together with a short leash. The dog is barking and the cat is meowing, which represents Bartholomew: bark/meow—Bartholomew.

Standing next to this scene is a man vigorously shaking his head in disbelief. He cannot imagine that such a scene is occurring. That, of course, is Thomas, the person who doubted that Jesus was raised from the dead.

Thomas has his money in a money pouch high up on his body. As he is shaking his head, the pouch is moving and the money is spilling out from a small hole in the bottom of the pouch and falling to the ground next to where he is standing. Sitting on the ground next to him and picking up the money with a certain amount of zest and counting it carefully is Matthew, the tax collector. You can embellish this picture by portraying Matthew using a calculator to add up all that he is collecting.

Thomas and Matthew are in an alfalfa field, and sitting next to Matthew is a rather large blue jay. This bird is watching Matthew pick up the coins as they fall. (Picture this jay doing this for some time, with his neck rotating dramatically up and down and back and forth.) The jay in the alfalfa field pictures James, son of Alphaeus.

The bird is so wrapped up in what he is watching that he does not notice a person walking rapidly right toward him. This fast-moving person is dressed in a zoot suit with bright, broad pin stripes down a very wide-lapelled jacket and suspenders underneath. He is walking with a "Keep on Trucking" kind of gait, and it is his foot extended in front that is ready to step on the blue jay's tail. This fellow is really into fads, and fad rhymes with Thad for Thaddaeus (who in some listings of the disciples is also called Judas, son of James).

With the hand that is not hooked into his suspenders, Thad is giving a good stiff arm to another fellow who is very zealously pursuing him. This man's eyes are bulging out with a Simon Legree look, his mustache is twirled at the end and he is reaching out after Thaddaeus. This person is Simon the Zealot (also called Simon the Cananean).

In his other hand Simon the Zealot has a very large rope—almost like one that would be used to dock a ship. The other end of this rope is tied in a hangman's noose around the neck of a man who is standing nearby with both arms down and his head remorsefully bowed down. This is, of course, Judas Iscariot, who hanged himself.

After you have gone through these illustrations once and described them as vividly and with as much action as possible, ask if someone else in the room could go through the very same illustrations, naming the associated disciples. Then ask for another. You will discover that most of the people in the room will now be able to name all of the disciples. In other words, our memories really are quite capable, but we do not always know how to harness them.

Ⓥ If time allows, next have them turn to Matthew 5:13 in their Bibles. Brainstorm with them some other interpretations of the word "salt" in this verse beyond those given in *Managing Yourself,* page 147. One use of salt is to preserve meat. In a similar fashion, Christians, in a sense, are to preserve society. Various salts are needed minerals in the body. Salt is used in certain sacrifices. Salt is a symbol of a covenant (see II Chronicles 14:5; Leviticus 2:13; Numbers 18:19; II Kings 2:20-21). Salt is an irritant to wounds, but performs the role of sterilizing the wound. (As you do your brainstorming, you will probably be wise to use your chalk board or other visual aid to stimulate further ideas to come from the group. Just write down each idea as a person shares it and thank him for it without evaluation or criticism.)

In all these cases there are parallel interpretations that describe the roles that Christians could or should play in the lives of the people around them. Interpreting and applying these analogies are examples to the group of "processing" kinds of mental activities at which they can become proficient.

5 min. ⓥ Indicate to the group that you will now be talking about the "sending" part of the mental area of life. Have the group turn in their Bibles to Proverbs 15 and once again brainstorm to see how many different specific ideas on how we should communicate can be gleaned from this Scripture passage in just a few minutes. Some, for instance, include: gentle speaking is more productive than words spoken in anger; fools despise instruction, but the wise accept it; wise counsel brings rich rewards, etc.

In "sending," as with "receiving" and "processing," the members of the group can learn additional detail in the text and in the "For Your Further Study" materials. The point of the exercises in this session is to show them that they can and should make better use of their God-given minds.

5 min. Now have each of them determine the one greatest need for improvement in the mental area of life. Next, encourage each of them to think of the one best way he or she

could make progress toward achieving his or her objective—just as he or she has learned to do in an earlier session.

1 min. Ask them to read Chapter 13 for the next session. Encourage the group members to place their personal application notes from this session in their planning/scheduling notebooks or files. Close in prayer.

ADDITIONAL MATERIAL (Supplementary Content Ideas):

1. Study some of the Scriptures mentioned in Chapter 12 that have not previously been covered in the session outline above. For example, Romans 1:28-31 gives a good illustration of how the mind can be misused by a person and can be instrumental in his or her spiritual downfall.

2. Have some members of the group give special reports on particular techniques they have learned that have improved the mental area of their lives. For example, if they have learned to "speed read," have one of them give a 5-10 minute lesson on speed reading to the group. Ask them to use the guidelines indicated on pages 150 and 151 as they prepare for the lessons. Ask them at the end to comment on how they were helped, particularly by the preparation guidelines given on page 151.

3. Do one or more of several additional exercises that can illustrate what the mind is capable of in various areas. For example, have one person mention a particular problem of which he or she is aware. Ask what he or she thinks is the cause of the problem. Listen to the answer, then ask what seems to be causing the "cause." Keep asking the same question, until you seem to have plumbed the depth of the true, ultimate problem. The use of this technique is a useful problem-solving technique that enables a person to be more sure to solve the problem instead of just the symptom.

4. Read John 13:34,35 to the group and ask the group for some ways in which love is important as a part of our communication to others. Then ask for some ways love

can be shown to people when communicating. (For example, listening can be a very important tool, not only in receiving, but also in sending the message that we care.)

5. Have two people in the group play the role of two people who are having real problems communicating. One person is trying to make a particular point clear and the other person consistently misses it. Have the group comment on what the sender and the receiver could do to improve.

6. Have the members of the group memorize Proverbs 23:7.

Session Eight

THE PHYSICAL AREA OF LIFE (Chapter 13)

SESSION OBJECTIVES (For Each Group Member):

1. To understand some simple biblical principles which help a person maximize his or her physical self to the glory of God.
2. To select one specific area of needed improvement and plan to implement it.

LEADER PREPARATION:

- ☐ Pray for God's wisdom in your preparation.
- ☐ Read and meditate on the session objectives listed above.
- ☐ Read carefully Chapter 13 of *Managing Yourself* at least once.
- ☐ Read the rest of this session's material several times.
- ☐ Determine what overall customizing of the material you plan to do.
- ☐ Prepare for personal illustrations, examples and group member assignments.
- ☐ Arrange for final needs related to facilities, visuals, refreshments, etc.

☐ Pray that special needs will be met in the lives of the group members and that the session objectives will be accomplished.

(See the Introduction section of this *Leader's Guide* for more detail.)

CONDUCTING THE SESSION:

1 min.   Start with prayer. Pray for the members of the group, that, to the extent they are neglecting this area of their lives, they will see the simplicity and the practicality of God's Word on this subject and that they will be motivated to take action.

8 min.   Read the session objectives, or at least introduce the subject to be covered by reading the title of the *Managing Yourself* chapter covered by this session.

Read the following parable to the group:

Once upon a time there was a gracious king who gave one of his servants a young colt out of the royal stables. This colt came from the finest of the stallions and mares owned by the king. The king indicated to his servant that he could use the colt as his horse as long as he would take good care of it. Initially the servant was very careful in raising the colt, but after a while he found he was interested in many other things besides his colt. He did want to make sure the colt didn't starve, so he gave him plenty of feed. But he didn't pay much attention to make sure the colt had the right amount and the right kinds of feed. He kept the colt in a small 10-foot-by-10-foot stall except when, from time to time, he needed to use him in various ways. This was only a couple of times a month, though. He never seemed to find time to brush the colt and wash him and do the various other things that would make him look good. He also didn't seem to notice when the colt seemed to develop sores or other physical problems. Finally, one day, the king came to visit his servant and asked to see the colt. The servant very sheepishly brought the king over to the stall where he kept the colt. The king looked at the colt and discovered what bad condition he was in and turned to his servant and said, "_____."

The question you now put to the group is, "What do you think the king should say to his servant?" Needless to say, the discussion will center on what a sloppy job the servant has done with the colt he was given, and how he really had not fulfilled the initial agreement he made with the king. After a brief discussion, ask, "What are some comparisons between the elements of this story and the gift of a body God has given to each of us?" The discussion then should go along the lines of the fact that God gives us our bodies in the first place when we are born. Then, when we become Christians He gives them to us again for our use—even though we are not our own but are bought with a price (I Corinthians 6:19,20). This compares to the colt that was given to the servant in the story.

If needed to make the point, ask the group, "What are some things God should say to us if we don't take good care of the bodies we have been give to use?" Of course, the group members will draw the connection between the parable and your question and answer that God should scold us, and, if He is very gracious, give us a little longer to shape up. If we don't, He may well allow us to lose the use of some of our physical gifts in the form of sickness or in other ways.

14 min. Ⓥ Highlight the first four principles from Chapter 13 of the text. You can do this either by lecturing without the group referencing their books or by allowing them to follow along in their books as you speak. Then ask the following three discussion questions:

1. What kinds of experiences have you had in which your reactions to bad circumstances actually made you ill? Their answers will vary, of course, according to experience.

2. What are some ways feelings of guilt could affect you physically? In this case they will also offer

various of their own thoughts. Be sure to include in the discussion the two passages mentioned in Chapter 13, Psalm 32:3,4 and Psalm 38:3-8.

3. Read Proverbs 17:22 to them: "A joyful heart is good medicine but a broken spirit dries up the bones." Ask them to think of some ways in which joy could actually cause physical benefits. Ask them if they have ever had that experience. Again, their answers will vary according to their experience.

13 min. Ⓥ Highlight principles 5-10 in Chapter 13. Once again, either lecture or have them follow with you in the book. (If you do lecture, it would probably be wise in this and the previous point to have the 10 principles written on your visual before the start of the session.)

Next, lead a discussion by asking the following series of questions, spending appropriate time with each question's answers:

What do you think of these ideas? Do they seem reasonable and practical? What are some ways you have discovered some of them yourself? What have you discovered specifically along these lines? The questions will generate various answers according to their experiences. The main point that you want them to see is that these are not radical or highly unusual principles. In many if not most cases the group members may have discovered the same principles themselves, though they may not have arranged them systematically in their own minds.

Next, have them use a piece of note paper for the following quiz (which will be self-graded):

1. How long has it been since you last had a physical check-up?
2. How well do you feel you control the quality of the food you eat? (For example, along the lines of the four points mentioned on pages 159-160.)
3. Would you say you eat too much food?
4. Would you say you weigh more than you should?

5. Do you feel you exercise as much as you should?

6. Do you feel you are able to sleep and otherwise rest as much as you should?

7. In the right sense of the word, are you proud of your appearance, as it relates to neatness and the care you give to your appearance?

At the end of the quiz, comment to the group that no doubt everybody came up with some areas where he or she could see the need for specific improvement.

8 min. Have the group members each select one particular area where they feel they need the most improvement. Ask them to think through an appropriate program to accomplish that improvement. Have them focus on what positive benefits this improvement would bring. Ask what barriers have historically prevented their accomplishing the desired results. For example, if they have always been troubled by dieting, what will they now do differently that will enable them to maintain a more realistic weight maintenance program?

1 min. Ask them to read Chapter 14 for the next session. Encourage the group members to place their personal application notes from this session in their planning/scheduling notebooks or files. Close in prayer.

ADDITIONAL MATERIAL (Supplementary Content Ideas):

1. Have the members in the group study some of the Scriptures mentioned in Chapter 13 that have not been featured in the above session. For example, Daniel 1 gives a very interesting passage for study. It highlights not only the physical benefits of not eating the wrong foods, but also the physical benefits of obeying God. I Kings 5-7 also gives evidence of the great majesty and beauty of the temple which Solomon built for the Lord and what great attention to detail went into its construction. This detail could be used as a symbolic comparison to the care and attention God has put into our bodies—which are also temples of God—and how much

care and attention they therefore deserve. As in previous Scripture studies, the passages can be read and discussed with the whole group or assigned to subgroups, which could then report to the larger group. The assignment could also simply be given for personal study.

2. The following is a physical program that one person outlined for me: He exercises mainly by running for one hour each morning and by working out with weights—doing presses and other forms of exercise—for one hour in the late afternoon. In addition to his normal food, he drinks three special milkshakes every day made up primarily of protein powder. Read this special physical program to the group and ask if this is one that would suit the average person. What are some reasons it would or would not? Who are some people for whom it may be good? (Obviously the example given is for an athlete in training who is doing things far in excess of what the average person needs to do.)

3. According to the interests of individuals in the group, have some of them do special reports on how to achieve progress in such areas as exercise, nutrition and reduction of anxiety. In the "For Your Further Study" section on Page 163 you will find some good books to help with these assignments.

4. Another alternative is for you to read passages from some of the above-mentioned books that cover ways to improve one's health. Lead a discussion on the specifics mentioned. You can also assign some of these passages to subgroups for discussion, to be later referred to the entire group for further discussion.

5. Have the members of the group memorize I Corinthians 6:19,20.

Session Nine

THE SOCIAL AREA OF LIFE (Chapter 14)

SESSION OBJECTIVES (For Each Group Member):

1. To understand the special importance of the social area of life to a Christian.
2. To discern the different levels of the social relationships of life and to learn how a person can receive from and contribute to others in each level.
3. To select at least one specific area of needed improvement and plan to implement it.

LEADER PREPARATION:

☐ Pray for God's wisdom in your preparation.

☐ Read and meditate on the session objectives listed above.

☐ Read carefully Chapter 14 of *Managing Yourself* at least once.

☐ Read the rest of this session's material several times.

☐ Determine what overall customizing of the material you plan to do.

☐ Prepare for personal illustrations, examples and group member assignments.

☐ Arrange for final needs related to facilities, visuals, refreshments, etc.

☐ Pray that special needs will be met in the lives of the group members and that the session objectives will be accomplished.

(See the Introduction section of this *Leader's Guide* for more detail.)

CONDUCTING THE SESSION:

1 min.
Start with prayer. Pray that God will show each group member the importance of the social area of life and the need to fit it into God's plan for his or her life. Also pray that he or she will come up with one specific area of improvement and seek to implement it as soon as possible.

5 min.
Have the group share their progress in the specific areas of improvement they selected for themselves in the previous three sessions. If you are aware of a particularly notable area of improvement in any one person's life, you might call on him or her. Probe for how this person has succeeded in his planning and implementing. If someone has run into a problem, ask the group what solution they would suggest.

5 min.
Read the session objectives, or at least introduce the subject to be covered by reading the title of the *Managing Yourself* chapter covered by this session.

Ⓥ
Ask the group to list some reasons the social area of life is important for a Christian and lead the group in discussing the question. You may wish to use your visual to acknowledge various points made. The following are some answers that might be given:

- The example of Jesus growing socially (Luke 2:52).
- The encouragement of Scripture to stimulate one another to love and good deeds and to fellowship together (Hebrews 10:24,25).
- The encouragement of Scripture to relate and compare fellowship with God and fellowship with man (I John 1:3).
- The significant influence of others on us for good or for bad (Hebrews 10:24; Hebrews 3:13; Proverbs 1:15; Psalm 1:1).

2 min.
Highlight the four levels of social relationships found on page 167 of *Managing Yourself.* Mention that you will be

Ⓥ deferring the discussion of family relationships to a later chapter and that this session will concentrate on the other three levels.

16 min. Have the group turn to page 168 in *Managing Yourself.* Indicate that you will be "workshopping" the section, "Close Friends." "Workshopping" means they will be asking themselves the questions asked in the book and trying to write down some answers that would make sense for their own situations. Give the group direction by asking such key questions as: "Who are your close friends?" After they have written a few names, ask: "Is everyone on this list living before God as he should? Or, are some living in such a way as to hinder their own Christian growth?" We are told in Scripture, in Mark 3:13, that Jesus carefully selected the people with whom He would be close friends. In their list most people will come across some within their close circle who are not entirely a blessing to them or vice versa. Ask them to think of at least one example of this before asking them the following questions (which are similar to those in *Managing Yourself*):

1. What are some ways you can minister to that person?
2. In what ways can you communicate to him how important it is for you to glorify God?
3. If God should lead you to "cool" this relationship, in what ways could you continue demonstrating your love to that person?
4. What are some ways in which you can continue with some form of relationship with that person and at the same time minimize the negative impact that person might be having in your life?

Next, have them list some needs in their lives and then list a close friend with strengths in those areas. If no person they are close to fits that category, have them think of someone else who is a potential close friend who does.

Now, reverse the proposition and name a friend whose greatest need is one you can meet through one of your areas of strength. Challenge each group member to determine how to take the initiative to cultivate a relationship with that person. Have them consider the ideas suggested on page 169.

Have a time of silent prayer during which each person will be praying specifically for their close friends which they have listed in this part of the session. Have them pray that God will really lead them to some specific opportunities that will result in a mutual upbuilding. Close the time of silent prayer by your praying aloud.

10 min. In this section of the session you will be covering the "Acquaintances and Other Friends" section of the chapter. Ask the group to think of a few specific people whom they know but *do not* consider as close friends and who could minister to them in an area of need or vice versa. This is normally a substantially larger group of people from which to pick and, therefore, it contains substantially more opportunities for ministry and blessing to occur. Have them quickly brainstorm and write down some ways that the impact of their life on this other person (or vice versa) could be maximized. Have them think of specific situations and conversations they could use to enhance a "fringe" friendship. For example, if the person in question likes to go bowling, this could be a natural opportunity for a beginning. During or after the game they could converse on subjects of mutual help and interest. Some people's greatest need is simply to let loose and have fun. The reverse is true with others. After the group has thought creatively for several minutes, have them check their lists and select the number one activities or approaches they could use to begin relationships in the near future with the persons they have listed.

5 min. Ⓥ As a final question for your session, ask, "How should we as Christians treat total strangers?" Encourage them to think of any scriptural obligations they might have to virtually everybody (such as Matthew 25:34-45 and Matthew 28:18-20). Are there realistic limits to what we can do for total strangers? If so, what are some of those limits? In what ways does a relationship with a stranger affect our relationships with those who have become our close friends? The discussion will probably go along the lines that we do have significant obligations even to total strangers, but at the same time God has made us responsible for certain special relationships that possess even greater obligations, such as family relationships (I Timothy 5:8).

1 min.   Ask the group to read Chapter 15 for the next session. Encourage the group members to place their personal application notes from this session in their planning/scheduling notebooks or files. Close in prayer.

ADDITIONAL MATERIAL (Supplementary Content Ideas):

1. Have the members in the group study one or more of the Scriptures mentioned in Chapter 14 that have not been featured above. For example, Philippians 2:1-4 is a good passage to study on the role of humility in foregoing one's rights in favor of the rights of others in order to have fellowship. It also points out the end product of unity that should come from the right kind of fellowship.

2. Have someone do a word study on the Greek word "*koinonia*," which means "fellowship" or "to have in common." You could also have him or her study the use of that word in I John 1. Have him or her report the findings to the group.

3. Give a specific homework assignment to the group to have one of their friends over for a meal or dessert. Have them aggressively seek to minister to or be ministered to by this person. Have them prepare specific questions to ask. Pray for each other in preparation for this time. Have them report back to the group next session.

4. Have the members of the group make a list of non-Christians with whom they are acquainted. Have them select the best person with whom to share the message of Christ's love and forgiveness. Train them in knowing how to share their testimony of how they came to Christ and to share the message of the gospel with this person. Pray for each other in preparation for the wit-

nessing situation. Have each member of the group report back on how it went.

5. Role play various ministering situations. If your group members are not well trained in social skills—such as conducting a conversation around the dinner table—then it would be good to give some sample conversations for them to see in action and then to discuss. Illustrate one or two "icebreaker techniques" that can be used to stimulate conversation. Included could be: asking people what they like to do and how they got started doing that particular thing (or occupation). Ask how they met their respective spouses. Teach how to use follow-up questions that will keep people interested and comfortable as you continue to share the gospel.

6. Have each member of the group memorize Hebrews 10:24,25.

Session Ten

THE VOCATIONAL AREA OF LIFE (Chapter 15)

SESSION OBJECTIVES (For Each Group Member):
1. To understand how important it is for Christians to pray and plan for the vocational area of their lives.
2. To know the five basic aspects of the vocational area of life.
3. To select at least one specific area of needed improvement and plan to implement it.

LEADER PREPARATION:
- ☐ Pray for God's wisdom in your preparation.
- ☐ Read and meditate on the session objectives listed above.
- ☐ Read carefully Chapter 15 of *Managing Yourself* at least once.
- ☐ Read the rest of this session's material several times.
- ☐ Determine what overall customizing of the material you plan to do.
- ☐ Prepare for personal illustrations, examples and group member assignments.
- ☐ Arrange for final needs related to facilities, visuals, refreshments, etc.
- ☐ Pray that special needs will be met in the lives of the group members and that the session objectives will be accomplished.

(See the Inroduction section of this Leader's Guide for more detail.)

CONDUCTING THE SESSION:

1
min.
Start with prayer. Pray for wisdom, especially in the vocational area of life, for the members of your group. There is a good chance that they will not have often thought of this area. Also pray that there will be a very meaningful application in their lives as a result of this session.

2
min.
Ask someone in the group if he or she would like to share how he or she is doing in the implementation of the application from the last session. If more people have interesting stories to share, you may want to allocate more time to this.

5
min.
Read the session objectives, or at least introduce the subject to be covered by reading the title of the *Managing Yourself* chapter covered by this session.

Ⓥ
Ask the group the question, "What are some reasons that it is important for Christians to think about the vocational area of life?" The answers will probably include comments such as, "That's an area of life we take for granted, even though much of our time is invested in it. Yet, as a Christian concerned with stewardship, one should prayerfully consider how we spend such large blocks of time." If time allows you might also ask, "Do you think it is scriptural to work?" The answers you receive might well follow the lines of some of the Scriptures listed on page 176 of *Managing Yourself*. Depending on the group, you may get some very interesting conversation in answer to that question.

1
min.
Ⓥ
List the five areas of emphasis in the vocational area of life as covered in Chapter 15: calling, training, apprenticeship, producing and growing. (You might want to have your visual aid prepared with these five areas already written.) Give the group a one-sentence definition or description for each area. (See the bottom of page 176 and the top of page 177 for these.) The following five parts of this session relate respectively to each of these five parts of the vocational area of life.

7 min. CALLING—Have each member of the group write down what he believes to be his calling in life. If the group has done a thorough job of planning in the earlier sessions, this may amount to a brief review. However, if much of their planning has had to do with improvements in personal areas of life as opposed to vocation, this may take longer. If this is the case, you may want to spend some time reviewing the main points from Session 5 concerning Chapter 9. You will then need to adjust your time schedule for this session. In any case, you should be sure that your group members have at least a fair idea of what God has called them to do vocationally.

Before going further, be sure your homemakers sense that the vocation of being a wife and mother is a very important one. Discussing Proverbs 31:10-31 and other related Scriptures should help.

Ⓥ If your group members feel quite clear about their calling already, you might ask, "What are some reasons it is important to know your calling?" The answers should include:

- It represents the greatest allocation of our waking hours.
- It potentially consumes the majority of our creativity.

7 min. TRAINING—Ask the group, as it relates to their respective callings, "What would you consider as good training for your calling?" Have them write what they think their training should be. Then ask if they feel well trained by the standards they have just written. If the answer is no, ask "In what ways can you become better trained for your vocation?" (The two paragraphs under "Consider Your Options" on page 179 will help you to discuss this area at greater length if you feel the need to do so.) Some members of the group may find that they really are not very well trained for what they are called to do. Some members of the group might help such persons think of creative ways to be realistic about how to solve their training

problems. If the group consists mainly of people doing the same sort of thing and if you can afford to spend more time on this, it would be particularly helpful to discuss specific training opportunities that exist for people with their vocation. If you are aware of the vocations of various people in the group and can do so, you might want to be prepared to suggest some books or tapes or other training program information that you are able to find.

As a final question related to training, ask the group to think of strengths they possess in terms of experience and training that can be cornerstones on which they can build their careers. Though some people do not have all the training they need, they could very well be trained in a particular aspect of their vocation. If so, some additional training might well qualify them more than they might have initially believed. The point of this question is to give them some optimism and confidence about the future.

7 min. APPRENTICESHIP—Ask the group if they can recall and tell the group about some particularly helpful apprenticeship situations that they have experienced. (You might relate an experience of your own in which an expert in some field taught you how to do something quite well. Our youth experiences are often full of such examples: someone showing us how to fish, how to tie knots, how to play ball. We each will have examples from our adult lives as well.) The main point of this discussion is to enable people to see how much they have been helped by the technique of apprenticeship.

As an option to the above you could start this session with a brief role play with one person observing another perform a significant activity with great skill. The observer is unimpressed, apparently disinterested to the point of appearing unteachable. Ask the group what they think of the observer's attitude. The discussion should relate to the observer's apparent unwillingness to learn from another, which will certainly adversely affect his ability to grow. Dr. Howard Hendricks says that a man's teachability is the key to his future growth.

After either option, have people write down how they could benefit right now in their jobs from some kind of an apprenticeship relationship. Have them try to be as specific as possible. Perhaps they can even write the names of the people and the particular subjects they would like to learn from those people. They might even think of when they would be able to approach these people concerning this.

7 min. (V) PRODUCING—Ask the group if they think it's a good idea to work hard, and if so, ask them to list some reasons. You might generate some very interesting discussion here. But I suspect that, having read some of the Scriptures in the chapter, some will say it honors God the most, not only when we work hard, but when we do things in an excellent fashion.

If time allows, have someone read Colossians 3:23,24 and I Corinthians 10:31. Ask the group how they think these verses apply to the vocational situations they face. The answers should include the concept of doing things in an excellent fashion in order to bring maximum glory to God.

If time allows, next read Galatians 5:13, particularly emphasizing how we are to serve one another through love. Ask the group to think of some way that relates to the job situation. The answers should include the concept of being a servant and being loving to those around us at work as well as at home.

Have them think of the top priority idea that they could apply out of this area to their own lives and have them write it down.

7 min. GROWING—Ask the group the following question, "What are some causes of growth in our personal capabilities?" The answers here should eventually come to the point that, when we are stretched in our work loads, when we have to face and solve problems, when we are given additional responsibilities, we are forced to grow.

Ask the group to think of and write down at least one specific way they could plan toward growing, personally, in their vocations. They should write down some potentially stretching situations that could help them grow.

If time allows, ask, "What are some ways in which we can cope with the pressures that come from these stretching experiences along with the other experiences that we face day to day?" The answer here should include the concept of learning to like what we must do. For some background here, turn back to pages 52-54 and you will see more Scripture and other reasoning related to how to be motivated. In the same chapter you also see some information related to a second point that should come out in the discussion: to be at peace despite circumstances. Another possible answer is to learn to enjoy a close relationship with God. As we walk closely with Him we receive power to handle pressure.

1 min. Ask the class to read Chapter 16 for the next session. Encourage the group members to place their personal application notes from this session in their planning/scheduling notebooks or files. Close in prayer.

ADDITIONAL MATERIAL (Supplementary Content Ideas):

1. Particularly in this session, you may want to look for ways to adapt to the special needs of your group. For example, you might invite a vocational counselor to talk to the group about choosing the right calling in line with their particular skills and interests. You could ask a Christian worker to present some of the opportunities in full-time Christian work. If your group is made up of people just out of school and into their first or second job, perhaps the emphasis should be on additional training. If such is the case you might invite a training or personnel director to speak on career planning, training and personal development. Or you could simply adjust the schedule to spend more time on one or more of the items being discussed.

2. Have the group study some of the Scriptures that were not studied in the session outline. For example, Romans 8:28-30 and Proverbs 31:10-31. Ask the group what they learned from these passages concerning the vocational area of life.

3. Ask the group for some ways that excellence and perfectionism differ. Then ask them for some ways this relates to the Christian functioning in his vocation.

4. Have each member in the group prepare a report concerning his or her calling, current level of training and plans for further training. Such an assignment would cause each member to do more thinking and detailed planning concerning his or her vocational area.

5. If the group consists primarily of homemakers, have them discuss how they could use some of their areas of strength to help those with areas of need. Consider a panel of homemaking experts to speak to the group.

6. Have each member of the group memorize Colossians 3:23,24.

Session Eleven

THE FINANCIAL AREA OF LIFE (Chapter 16)

SESSION OBJECTIVES (For Each Group Member):

1. To understand the importance of the financial area of life to the Christian.
2. To know the key biblical principles of this area of life.
3. To select one specific area of needed improvement and plan to implement it.

LEADER PREPARATION:

☐ Pray for God's wisdom in your preparation.

☐ Read and meditate on the session objectives listed above.

☐ Read carefully Chapter 16 of *Managing Yourself* at least once.

☐ Read the rest of this session's material several times.

☐ Determine what overall customizing of the material you plan to do.

☐ Prepare for personal illustrations, examples and group member assignments.

☐ Arrange for final needs related to facilities, visuals, refreshments, etc.

☐ Pray that special needs will be met in the lives of the group members and that the session objectives will be accomplished.

(See the Introduction section of this *Leader's Guide* for more detail.)

CONDUCTING THE SESSION:

1 min. Start with prayer. Pray for the group's understanding of the key principles of finances as presented in the Bible. Pray that they will see how this awareness can reflect the priority God has in their lives and how the proper financial priorities can spare them great anguish.

1 min. Review briefly the five aspects of the vocational area of life covered last week.

8 min. Read the session objectives, or at least introduce the subject to be covered by reading the title of the *Managing Yourself* chapter covered by this session.

Ⓥ Ask the group the following three questions with time for answers and discussion in between:

1. What are some benefits of income and wealth? Possible answers:
   - Provision for needs
   - Enjoyment
   - Opportunity to be a blessing to others by giving

2. What can be some dangers of too much wealth? Possible answers:
   - Stopping trusting God
   - Starting loving the world

3. Does that mean that all Christians should be poor? If not, for what reasons? Possible answers:
   - No, but the more of the world's goods we possess, the more there is for us to handle and to be good stewards over.
   - No, but Christians need to be sure they don't put the things of this world ahead of God in their lives.

Announce that you will now proceed to discuss each of the three main sections of the chapter concerning the perspective, priorities and possibilities involved in the financial area of life.

8 min. PERSPECTIVE—Make the statement that God created all things as documented in Genesis 1 and 2 and Revelation

Ⓥ 4:11. Read Philippians 4:19. Ask the group what they think that means. The discussion should focus on the fact that God really does promise to meet all of our needs. Ask, "To whom should we go when we have a need?" The answer, of course, is God. Then ask, "What are some things we usually do when we are short financially?" The answers probably will include:

- Get upset (which amounts to indicating trust in ourselves, and we are not worthy).
- Use our balance plus or charge card (which indicates our trust in the banker to meet our needs).
- Ask for a raise from our employer (which indicates a trust in him).

Next ask, "What are some things we can meaningfully do to trust God?" Let the group discuss this briefly. The essence of the answer should be that we should pray and ask Him for His provision, which may include going without what we initially thought we needed.

If time allows, toss this statement-question to the group: "It has been said that one of the leading causes of anxiety is having our rights (such as time, respect, possessions) violated. Would you agree or disagree and what are some reasons for your opinion?" The discussion here should come to grips with the fact that when we feel we "possess" things, we have a tendency to feel hurt or disenfranchised when we lose them. This feeling of possession by the Christian demonstrates a failure to recognize that we are stewards and not owners of whatever might be in our charge while we are here on earth. Some background on this can be found on pages 191 and 192 in Chapter 16.

8 min. Ⓥ PRIORITY—List and briefly define the four priority areas of disbursement for the Christian: God's work, your family, your government and your debts. Material for this can be found on pages 192-194 of Chapter 16. Then ask the group for some reasons a Christian should give priority to giving to God's work. Possible answers include:

- An acknowledgement of God as the source of our incomes.

- A demonstration of the preeminence of God in our lives. We are commanded to give Him the first part of our produce (Proverbs 3:9,10).

Next, ask the question, "What are some problems of being in debt?" Possible answers include:

- We become the lender's slave (Proverbs 22:7).
- It is a presumption on God that He will not only meet our future needs, but also enough to pay back our debts.
- It becomes a possible distraction in our work for the Lord as an entanglement in the affairs of everyday life (II Timothy 2:4).

If time allows, continue in this same vein and ask for some reasons that it is important to provide for your family and to pay your taxes.

8 min. POSSIBILITIES—List and briefly define the four possible places where we can allocate our surplus income: giving more to God's work, giving to the needs of others, investing for future needs and spending on desires.

Next, read II Corinthians 8:12-15 and ask the group, "What are some things this Scripture tells us about how God often meets needs?" The answers should include the fact that God uses a surplus in the hands of some of His children to help meet the needs of others of His children. Next, ask the group, "What happens if we withhold a surplus designed to go to meet a specific need?" The answer, of course, is that the need goes potentially unmet.

Next ask, "Which of the four possible areas of allocating surplus do you think will normally get the surplus?" Most people would probably vote for the fourth possibility, that of spending on our desires, normally gets the surplus.

Ask, "For what reasons is this so?" And ask, "What are some ways we can avoid this automatic choice and make a real choice among all four possibilities?" The answers here are that we tend to spend money on things we enjoy, which includes a lot of items which could be classified as "desires." The only way to make a real choice is to go through a thoughtful and prayerful process, more slowly considering all four possibilities, thus giving the other three a greater chance of being selected.

5 min. Give the group the following quick quiz. Assure them that they will score their own quiz, but that the purpose is to reveal if there are some special needs in their lives for improvement in handling funds.

1. Do you *use* a written budget in your family?
2. Do you pray about what you buy before buying it particularly before major purchases?
3. Would you say you are an impulse buyer?
4. Do you know what you would cut out if your income were to go down (perhaps even due to inflation)?
5. Do you research and/or do comparative pricing on your purchases?
6. Do you tend to throw away items instead of repairing them?
7. Do you often use credit cards so you can buy something now even though you don't have the money now?

Ask the group to judge for themselves if they could afford to improve some of their financial habits.

5 min. Ask each person to select one area covered during this session that is most in need of improvement in his or her life. When each person has done so, ask the group members to brainstorm (on paper) some ways in which they can start implementing improvement in this highest priority need area. Have them each look over his or her list of program ideas and select the number one priority activity.

1 min. Ask the group to read Chapter 17 for next time. Encourage the group members to place their personal application notes from this session in their planning/scheduling notebooks or files. Close in prayer.

ADDITIONAL MATERIAL (Supplementary Content Ideas):

1. Have the group study and discuss some of the Scriptures that were not featured above, such as I Corinthians

9:6-14. Ask the group what each passage teaches about the financial area of life.

2. Arrange for a debate between two members of your group on whether or not taxes are to be paid even if waste and unfairness exist in government. They are to base their arguments for or against strictly on Scripture. Have the group discuss the subject after the debate.

3. Ask the group to define the word "needs" as it relates to food, clothing and shelter. What are some ways in which "need" differs from "desire" in each of these areas.

4. If the group normally sings as part of their time together, have them learn and sing the popular chorus "Seek Ye First" or some other song appropriate to this subject.

5. Discuss as a group what is a "reasonable" amount to invest for future needs and what would be "excessive."

6. Have each member of the group memorize Matthew 6:33.

Session Twelve

THE FAMILY AREA OF LIFE (Chapter 17)

SESSION OBJECTIVES (For Each Group Member):
1. To understand the importance of the family area of life for a Christian.
2. To discern the various decisions and functions a Christian must make in this area and to learn how to decide and do well in each area.
3. To select one specific area of needed improvement and plan to implement it.

LEADER PREPARATION:
- ☐ Pray for God's wisdom in your preparation.
- ☐ Read and meditate on the session objectives listed above.
- ☐ Read carefully Chapter 17 of *Managing Yourself* at least once.
- ☐ Read the rest of this session's material several times.
- ☐ Determine what overall customizing of the material you plan to do.
- ☐ Prepare for personal illustrations, examples and group member assignments.
- ☐ Arrange for final needs related to facilities, visuals, refreshments, etc.
- ☐ Pray that special needs will be met in the lives of the group members and that the session objectives will be accomplished.

(See the Introduction section of this Leader's Guide for more detail.)

CONDUCTING THE SESSION:

1 min. Start with prayer. Pray for the members of your group, that they will make good decisions in this area. Pray that they learn to demonstrate more love and become better communicators and team members in their marriages. Pray that they become better Christian examples to their spouses, children, and to those who observe their marriage.

2 min. Review what was covered last week in the financial area of life. Highlight the major points under perspective, priorities and possibilities and/or have one person share some progress that he or she has already seen in that area of his or her life or a commitment he or she has made.

3 min. Read the session objectives, or at least introduce the subject covered by reading the title of the *Managing Yourself* chapter covered by this session.

Ⓥ Ask the group, "What are some ways the family area of life is important?" Possible answers include:
- The need for an example to the children and others.
- A context for raising children.
- A source of happiness in life for husband and wife.
- Specific Scriptures group members may volunteer.

7 min. The following are two discussion questions related to whether or not to marry and whom to marry:
1. Make this statement, "She really deserves a husband." What are some things wrong with this statement? Possible answers include:
   - God is sufficient to meet any person's need.
   - We don't earn marriage.
   - Maybe God's plan for her is to be single.
   - It isn't good to put pressure on her directly or indirectly.
2. Make the statement, "They are so in love, it must be God's will for them to be married." Ask, "Are there some things wrong with this statement? If so, what?" Possible answers include:

- Love in the emotional sense is only a part of the commitment and the cause for marriage.
- The question still remains, is it truly God's will?
- Do they truly complement one another?
- Are they compatible in callings?
- Are they a spiritual match?
- Are they able to be close friends or are they just infatuated? (I am personally persuaded that adults should do less commenting along the lines of "Isn't that sweet?" and more commenting along the lines of "Are you sure?")

16 min. Ⓥ The following are some discussion questions related to developing the marriage relationship, loving your spouse and communicating with him or her:

1. Ask the members of the group to list five specific, deliberate, different actions they have taken to develop the relationship with their respective spouses since they've been married. Give them a couple of minutes. Then ask them if it was a problem for them to think of such actions. And, if such is the case, what can they now do to help develop their relationships? For example, they could plan definite times together, they could commit themselves to talking more to one another in the evening instead of watching TV and they could commit themselves to plan together as a couple.

2. Ask, "What are some ways we can show love for our spouses?" Possible answers include:
   - Doing things they appreciate. (Remember how you sought to do that when you were dating the person who is now your spouse.)
   - Saying "I love you" more often.
   - Picking up your socks in the bedroom, etc.
   - Having special times away together.
   - Being affectionate.
   - Implementing I Corinthians 13:4-8 in the relationship (e.g., being patient).

Ask your group to decide which of these things would be the highest priority for them to do now to show love for their spouses.

3. If time allows, conduct a role play with a husband and wife at the breakfast table. The husband has a newspaper in front of his face and a cup of coffee in his hand. He is paying no attention to his wife, simply grunting to her attempts at conversation, while reading and shifting the pages of the newspaper and sipping his coffee. The wife on the other hand, is talking constantly, hardly pausing for a response. At the end of the role play, ask the group what violations they observed relative to good communications. Possible answers include:

- The husband was not listening and understanding what was being said.

- The wife kept talking on.

- There was a noticeable lack of concern on the part of the husband for the wife.

- There was little effort on both people's parts to make themselves interesting and attractive to one another.

At this point lead them to the outline on pages 212 and 213 in *Managing Yourself* to see if some other suggestions by Howard Hendricks could be relevant to this couple. Finally, ask the group what one thing they feel they can do to communicate better with their spouses. Have them write down their ideas.

11 min. The following are some discussion questions related to whether or not to have children and how to raise them:

Ⓥ 1. Ask the group, "What are some pros and cons of having children?" Possible answers include:

- They are a gift of the Lord and a reward (Psalm 127:3).

- It is God's normal plan for continuing the human race.

- Christian marriages should offer the best setting in which children can be raised.
- On the other hand, children are a distraction from time and service that can be given to the Lord. I Corinthians 7 points this out relative to the marriage relationship.

If you have a group that's concerned with the issue of having or not having children, you could benefit by spending a good share of this session's discussion time on this question. Chapter 9 of *Managing Yourself* will help provide information on seeking God's will relative to children.

2. Ask the group to share ideas that they have on how to raise children. As the group leader, try to see that some of the 12 ways that Howard Hendricks mentions on pages 214 and 215 of *Managing Yourself* are represented in the discussion.

You might want to further stimulate the discussion by asking questions like, "What are some ways your family has fun together?" Or, "What are some ways you discipline your children to develop them?" At the end of this discussion, ask each member of the group to determine is his or her own number one improvement area in raising children.

4 min. Next, have each member of the group determine—from all the above ideas—his or her personal number one need for improvement in the family area. After each has done this, have each group member brainstorm (on paper) some program activities that will help meet that need. Then suggest that each of them select a number one activity to start working on.

1 min. Ask the group to read Chapters 7 and 18 for the next session. Encourage the group members to place their personal application notes from this session in their planning/scheduling notebooks or files. Ask each group member to review the notes he or she has accumulated in his or her planning/scheduling notebook or file and to bring his or her notebook or file to the next session.

ADDITIONAL MATERIALS (Supplementary Content Ideas):

1. In this particular session more than most of the sessions, you need to be careful to discern the needs of your group and customize the session accordingly. For example, if the group consists largely of unmarried people, you should place more emphasis on the single person's decisions relative to family. You can also guide them in discussing how they can help their parents fully implement the role of parents in their lives, and how they can honor their fathers and mothers. Another special content idea would be to use Howard Hendrick's tape titled "Marriage: Communication or Chaos?" (See publishing information on page 218.) You could also have one or two members of the group do a book report on some of the books mentioned on pages 217 and 218. You could invite a marriage counselor to speak to the group, following with a group discussion. A panel of outside "experts" might provide some fresh input.

2. Have the group study some of the Scriptures that have not already been studied in the above outline (e.g., Ephesians 5:22-25; I Corinthians 13:4-8). Probe for the practical implications of what these verses are saying relative to the marriage relationship.

3. If some of your people are doing very well in this area, have them give a report on their progress. For example, if some have succeeded in raising some outstanding children, the group might profit by having the parents tell how they did it.

4. This session lends itself to the use of role playing. A good one to consider could be a discussion between a young man and woman on a date concerning the subject of marriage. You could also role play other interaction situations. Note how this was handled with the breakfast table scene in the "Conducting the Session" section.

5. Have someone in the group prepare a 5-10 minute lesson for the group on how to disciple a person. Have the

group discuss how this procedure could be used with either a husband-wife or parent-child relationship.

6. Have each member of the group memorize I Corinthians 13:4-8 or I John 4:7.

Session Thirteen

PUTTING IT ALL TOGETHER AND KEEPING AT IT (Chapter 7)
CONCLUSION (Chapter 18)

SESSION OBJECTIVES (For Each Group Member):

1. To review the content covered in the previous 12 sessions.
2. To check the degree of understanding and implementation of the key concepts covered.
3. To become more motivated to keep implementing what has been learned.
4. To pray for one another.
5. To emphasize the priority of walking closely with God.

LEADER PREPARATION:

☐ Pray for God's wisdom in your preparation.
☐ Read and meditate on the session objectives listed above.
☐ Read carefully Chapters 7 and 18 of *Managing Yourself* at least once.
☐ Read the rest of this session's material several times.
☐ Determine what overall customizing of the material you plan to do.
☐ Prepare for personal illustrations, examples and group member assignments.
☐ Arrange for final needs related to facilities, visuals, refreshments, etc.

☐ Pray that special needs will be met in the lives of the group members and that the session objectives will be accomplished.

(See the Introduction section of this Leader's Guide for more detail.)

CONDUCTING THE SESSION:

1 min. Start with prayer. Pray for overall understanding of the content previously covered. Pray that each person will receive adequate motivation to follow through on what has been learned and what he or she has begun to implement in his or her life. Pray also for a special awareness of the importance of praying for one another and walking closely with God.

6 min. Read the session objectives, or at least introduce the subject to be covered by reading the titles of the *Managing Yourself* chapters covered by this session.

Ⓥ Give a brief general overview of what has been covered in the previous 12 sessions. Point out that there are two major sections to the content they have received, "The Basics of Managing Yourself" (Part I) and "Beyond the Basics: Tools for Further Effectiveness" (Part II). The Table of Contents is a good outline to use in your overview and I suggest you have your group turn to theirs as you use it to explain what was covered in Part I. Go through Part I chapter by chapter, with a one- or two-sentence summary of each. Do the same with Part II. Your intent is simply to call back to their minds what they have covered.

Next, highlight from Chapter 7 the concepts covered on pages 83-85 under the headings, "Harness Your Strengths" and "Compensate for Your Weaknesses." Point out that God has created them with particular strengths for which they have an obligation to be good stewards. He also allowed them to have certain weaknesses, some of which they are now developing or will develop in the future. However, these weaknesses can normally be recognized and in some measure compensated for in the meantime.

Encourage them to tell others about what they have learned. As they do so, they will more firmly implant these truths and concepts into their minds, and they will be a blessing to others as well. Perhaps some of them may even want to initiate a study group like this.

15 min. Ⓥ Indicate you are now going to review long-range planning, including their own implementation of it. Lead off by asking, "What are some reasons that it is important to plan long range?" The answer should relate to the need to know where we are going over a longer period of time which will enable us to be both efficient and effective. Ask how many have their long-range objectives written out and transferred to 3 x 5 cards along with their number one priority objectives for the next six to twelve months, and their number one program activities leading toward that number one objective for the next six to twelve months. Ask them if they still have these cards posted prominently where they can see them every day. Of those who keep their cards before them every day, ask how that has worked for them. Encourage them to share the method's effectiveness with the rest of the group.

Ask the group if each of them is ready to go on to a new number one priority objective for the next period of time. Many of them will already have made his or her existing objective a habit no longer requiring special emphasis. Ask the rest of the group what seems to be standing in the way of their implementation of their plans. (If time allows, offer suggestions yourself or allow members of the group to offer suggestions to help them break through their barriers.)

Next, have the group members open their planning/scheduling notebooks and files. Have them select from among the various ideas the top three priorities for them to consider implementing in the next six to twelve months. Have them turn to page 106 and, using that format, quickly come up with plans for their number one areas currently needing improvement. Given all of their experience in doing this kind of thing, I think in several minutes they can make significant progress.

Encourage the group members to keep their notebooks or files in a handy place where they are reminded to browse through their personal application ideas from time to time (at least every few months). When their current number one priority objectives basically become habits in their lives, they should replace those with new ones. They should update their 3 x 5 cards or replace them accordingly.

Encourage them to enter new ideas in their notebooks or files as they think of them. Urge them not to put those ideas on scraps of paper which they keep on the kitchen counter for awhile and then lose.

6 min. Ⓥ Briefly review the four steps of "How to Schedule Your Time" with the group (list activities, ask if assignable, assess priorities and schedule). Strongly emphasize the importance of keeping at the number one priority. Ask who has used these principles in their scheduling over the last few days. Ask one or more of those who have done so to share the benefits this practice has brought into their lives. Ask those who were not able to do so to share the reasons why it has proved difficult for them to implement this. Offer helpful suggestions yourself and let others in the group do the same.

Finally, if time allows, ask "What are some reasons that it is important, after all, for Christians to schedule their time?" Look for a couple of comments concerning the importance of making the most of that special resource God has given us.

6 min. Ⓥ Quickly review the four main principles involved in following your schedule (motivation, discipline, sensitivity and peace). Ask for volunteers from the group to share how one or more of these concepts has helped them.

Ask the group for some reasons that they think it's important to master these four principles in their lives. Look for answers along the lines of how it makes it easier to follow through when we are motivated; how we can't really grow if we aren't at least somewhat disciplined; how important it is to stay sensitive to God and to people lest we be doing things and coming across in a way that God never intended for us to do; and how important it is to experience peace in spite of the great pressures that we can sometimes feel.

If time allows, end this particular section with a few highlights from the section, "How to Keep from Stopping" on pages 81 and 82 in Chapter 7.

8 min. Have an extended time of prayer for one another in the areas expressed above. Praise God for progress in people's lives and for specifically answered prayer. Pray earnestly for remaining needs in people's lives, especially for ones who have tried and not yet succeeded. After you have closed in prayer, encourage them to keep on praying for one another in the future. You might suggest they even pair up one on one or by couples for that purpose.

3 min. Encourage the group members to place their personal application notes from this session in their planning/scheduling notebooks or files.

End this final session by drawing their attention back to the spiritual prerequisites for *Managing Yourself.* Remind them of how important it is to be assured that we do know God and are walking closely with Him, to be cleansed from the guilt of our sins and to be filled and empowered by Him. There is nothing more essential to managing yourself than this.

ADDITIONAL MATERIAL (Supplementary Content Ideas):

1. Have the group study the Scripture related to the parable of the talents (Matthew 25:14-30). Ask the group what insights they have gained into the subject of stewardship as they have been learning from this study over the last 12 sessions.

2. Have each member of the group prepare a short testimony in which he or she shares the most important

thing he or she received from this study and why.

3. Have the members of the group practice sharing some of the concepts they have learned in this study. For example, have them explain the "How to Schedule Your Time" outline with the person who is sitting next to them. Then reverse roles, and have the other person explain another outline. This will get them started passing the concepts on to other people and also learning the concepts better themselves.

4. Ask the group the question, "What will be the greatest hindrances for *you* in following through on what you have learned? How specifically will you seek to overcome these hindrances?"

5. Have the group members write 100-word statements on why it is important for them to manage themselves.

6. Have each group member memorize I Thessalonians 2:12, which highlights one of the main reasons to learn personal management.